The Economics of Technology Diffusion and Energy Efficiency

ADVANCES IN ECOLOGICAL ECONOMICS

Series Editor: Jeroen C.J.M. van den Bergh, *Professor of Environmental Economics, Free University, Amsterdam, The Netherlands*

Founding Editor: Robert Costanza, *Director, University of Maryland Institute for Ecological Economics and Professor, Center for Environmental and Estuarine Studies and Zoology Department, USA*

This important series makes a significant contribution to the development of the principles and practices of ecological economics, a field which has expanded dramatically in recent years. The series provides an invaluable forum for the publication of high quality work and shows how ecological economic analysis can make a contribution to understanding and resolving important problems.

The main emphasis of the series is on the development and application of new original ideas in ecological economics. International in its approach, it includes some of the best theoretical and empirical work in the field with contributions to fundamental principles, rigorous evaluations of existing concepts, historical surveys and future visions. It seeks to address some of the most important theoretical questions and gives policy solutions for the ecological problems confronting the global village as we move into the twenty-first century.

Titles in the series include:

A Theory of the Environment and Economic Systems
A Unified Framework for Ecological Economic Analysis and Decision-Support
Reinout Heijungs

Sustainability in Action
Sectoral and Regional Case Studies
Edited by Jörg Köhn, John Gowdy and Jan van der Straaten

Sustaining Agriculture and the Rural Environment
Governance, Policy and Multifunctionality
Edited by Floor Brouwer

The Economics of Technology Diffusion and Energy Efficiency
Peter Mulder

Time Strategies, Innovation and Environmental Policy
Edited by Christian Sartorius and Stefan Zundel

America's Changing Coasts
Private Rights and Public Trust
Edited by Diana M. Whitelaw and Gerald R. Visgilio

The Economics of Technology Diffusion and Energy Efficiency

Peter Mulder

Post-Doctoral Researcher, Department of Spatial Economics, Free University, Amsterdam, The Netherlands

ADVANCES IN ECOLOGICAL ECONOMICS

Edward Elgar
Cheltenham, UK • Northampton, MA, USA

Published by
Edward Elgar Publishing Limited
Glensanda House
Montpellier Parade
Cheltenham
Glos GL50 1UA
UK

Edward Elgar Publishing, Inc.
136 West Street
Suite 202
Northampton
Massachusetts 01060
USA

A catalogue record for this book
is available from the British Library

ISBN 1 84376 823 2

Printed and bound in Great Britain by MPG Books Ltd, Bodmin, Cornwall

Contents

Acknowledgements

This book was mainly written when I worked at the Institute for Environmental Studies (IVM) of the Free University in Amsterdam, while a substantial part of the research has been carried out at CPB Netherlands Bureau for Economic Policy Analysis in The Hague. I gratefully acknowledge access to their facilities, including generous provision of the data and software required for various chapters.

I am greatly indebted to Henri de Groot: without his numerous careful comments and useful suggestions this book would have looked much different, less interesting and it would not yet have been finished. He co-authored Chapters 2–6. Moreover, I am indebted to Marjan Hofkes and Daan van Soest for their contributions to Chapters 2 and 3 and Chapter 4 respectively.

I benefited greatly from comments on earlier drafts of various chapters by Frank den Butter and Peter Nijkamp as well as Jeroen van den Bergh, Kornelis Blok, Casper van Ewijk, Frank Geels, Reyer Gerlagh, Egbert Jongen, Richard Nahuis, Michiel de Nooij, Mark Lijesen, Ton Manders, Hein Mannaerts, Machiel Mulder, Sjak Smulders, Paul Tang, Harmen Verbruggen, Herman Vollebergh, Cees Withagen and Bob van der Zwaan. I thank Luke Brander for improving the linguistic quality of the book and Fons Groot for his help with computer issues. I owe Linda van de Kamp many thanks for her help with the final details, and much more.

The financial support of the Netherlands Organisation for Scientific Research (NWO), which made this book possible, is gratefully acknowledged. Finally, I would like to thank the editorial staff at Edward Elgar for their excellent support in the preparation of this book.

Peter Mulder

PART I

Introduction

1. Technological change, economic growth and energy use

1.1 INTRODUCTION

Energy is an essential factor that fuels economic growth and serves human well-being. World primary energy use has grown enormously since the middle of the nineteenth century due to unprecedented growth of population and technological revolutions in industry. This increase in the scale of energy demand comes at a certain price, including environmental externalities. Nowadays one of the most prominent environmental problems is the enhanced greenhouse effect, which is to a large extent caused by the cumulative impact of burning fossil fuels on an increasing scale, in order to meet the increasing energy demand. Currently most governments in the developed world strive explicitly for sustainable development, aiming to decouple economic growth and environmental pressure. Notwithstanding the need for renewable energy sources, this also asks for further improvements in energy efficiency. Technological change plays a crucial role in both developing alternative energy sources and realizing energy efficiency improvements, and hence in ameliorating the conflict between economic growth and environmental quality. At the same time it is known that diffusion of new technologies is a lengthy process and that many firms do not invest in best-practice technologies. These are the issues that set the stage for this book.

The aim of this book is to contribute to our understanding of the interplay between economic growth, energy use and technological change. The principal focus of this book is the adoption and diffusion of energy-saving technologies. This implies a choice to analyse technological change rather than economic growth, diffusion of technologies rather than their innovation, and energy use rather than other environmentally damaging activities. The first concern of this book is with the dynamic aspects of the choice of technology at the firm level and how this choice is affected by characteristics of technological change as well as by environmental policy. With this aim, we develop two small analytical models of technology adoption, inspired by recent developments in economic theorizing about technological change that stress the role of knowledge accumulation (learning), knowledge spillovers, uncertainty and irreversibility (path dependency). Does this view of technological change shed new light on the observed slow diffusion of new energy-saving technologies? What are the possible implications for the role of policies aiming to stimulate the adoption of energy-saving technologies?

The second concern of this book is with empirical regularities of productivity performance. We will explore long-run trends in energy and labour productivity performance across a range of OECD countries at a detailed sectoral level. To what extent has the pursuit of decoupling energy use and economic growth within the OECD been successful over the last decades? Why do productivity levels differ across countries and across sectors? For the purpose of answering these questions we will document stylized facts, decompose long-run trends and provide an extensive empirical convergence analysis of cross-country energy and labour productivity performance. The study concludes by integrating the insights from the small analytical models developed in this book, in an existing Dutch policy model of economy–energy interaction. The aim of this exercise is to investigate the possible relevance of these theoretical insights for applied economic policy analysis and to examine their impact on estimates of energy demand at a sectoral level. In the remainder of this chapter we briefly introduce the above-mentioned issues and the way they will be dealt with in the research reported in this book.

1.2 ECONOMIC GROWTH AND TECHNOLOGICAL CHANGE

Technological change is the main driving force behind economic growth. Although this message was already put forward by classical economists, most notably Schumpeter (1934), it was not until the 1950s that the crucial role of technological change in explaining patterns of economic growth attracted broad attention in mainstream economics. This development can be mainly attributed to the introduction of the neoclassical growth model, developed by Solow (1956) and Swan (1956), aiming to provide a theoretical framework for understanding geographical differences in patterns of growth of per capita output. In the model, aggregate output depends on capital, labour and labour-augmenting technological change. The model is an intellectual masterpiece, it fits the major stylized empirical facts (Kaldor 1961) and became the benchmark for what is now called the neoclassical theory of growth. However when Solow (1957) used the model for growth accounting it turned out that it was unable to explain growth rates of output on the basis of accumulation of physical inputs (capital and labour) alone. A large and persistently positive residual remained, which has become known as the Solow-residual. The remarkable conclusion was that something different from capital accumulation and an increasing labour force should be held responsible for most of the economic growth that has occurred. This 'something' was identified as 'total factor productivity growth' and has often been equated with 'technological change'. For this reason, the Solow-residual is often referred to as the 'measure of our ignorance' (Abramovitz 1986). Reducing this ignorance became the starting point of a quest for a better understanding of the process of technological change and nowadays many economists are still looking for missing pieces of that scientific Grail. Many things though have been learned in the meantime, and different schools of thought have contributed. An interpretative survey of this quest is provided in Chapter 2, with a focus on two recently emerged classes of models on economic growth and technological change. They are respectively labelled neoclassical endogenous growth models and evolutionary growth models. Both classes of models emerged out of dissatisfaction with the Solow–Swan growth model, in particular with its assumption that technological change is exogenous – suggesting that technological change falls like manna from heaven.

The recently developed class of neoclassical endogenous growth models moves beyond this unsatisfactory feature by endogenizing technological change. These models stress the fact that in the end, technological change is all about knowledge creation. In a growth perspective, this implies that economic growth is ultimately driven by accumulation of knowledge or human capital, thereby generating positive externalities (spillovers). The essential idea in the first endogenous growth models by Romer (1986) and Lucas (1988) is that knowledge is to be considered as a kind of renewable capital good, with capital being a broad concept including knowledge or human capital. Following this path-breaking research, many models have been developed that endogenize technological change, highlighting different dimensions of technological change and fields of application, including energy–economy–climate interactions (for example Goulder and Schneider 1999, Goulder and Mathai 2000).

The evolutionary growth models move beyond neoclassical endogenous growth models by criticizing not only some assumptions of the neoclassical growth models, but the whole neoclassical economic view on technological change as such. The foundations of the evolutionary economic view of technological change have been laid down in the seminal work of Nelson and Winter (1982), who argued that 'the weakness of the neoclassical theoretical structure is that it provides a grossly inadequate vehicle for analysing technical change. In particular, the orthodox formulation offers no possibility of reconciling analyses of growth undertaken at the level of the economy or the sector with what is known about the processes of technical change at the micro level' (Nelson and Winter 1982: 206). In sum, with the latter they meant that technological change is the result of search processes, which are inherently uncertain and guided by 'routines'. Therefore they analyse technological change in terms of strategies to optimize profits under the assumption of bounded rationality rather than to maximize profits under the assumptions of full rationality and perfect foresight. Nelson and Winter (1982) developed an evolutionary growth model along these lines, which they showed to be capable of reproducing the patterns of aggregate output, factor input and factor prices as addressed by Solow (1956) in his pioneering contribution to neoclassical growth theory. Following this work, many models have been developed, analysing technological change from an evolutionary point of view, emphasizing the role of uncertainty, learning, irreversibility and path dependency (see for example Dosi et al. 1988, and Freeman 1994 for a survey). Whatever one's opinion on these developments is, they meant a revival of the evolutionary tradition in economics, which goes back to Veblen (1898) and Schumpeter (1934), as a result of which evolutionary theories and models have become the most influential alternative to neoclassical theories and models on economic growth and technological change. As noted above, the current neoclassical view on technological change is not the same as it was in the days of Solow, and hence the question arises whether neoclassical economics is still 'a grossly inadequate vehicle' for analysing technical change. We deal with this question in Chapter 2, searching for the essential differences and similarities between the neoclassical and evolutionary economic view on technological change, both conceptually as well as in terms of modelling practices.

1.3 INNOVATION VERSUS DIFFUSION

In this book we adopt the widely used typology of technological change as has been provided by the Schumpeterian trilogy of invention (generation of ideas), innovation (development of those ideas through first marketing or use of a technology) and diffusion (spread of new technology across its potential market). However, as noted previously, the focus of this book is diffusion rather than innovation or invention. It is beyond doubt that the innovation of new technologies is at the very heart of processes of technological change. However the diffusion of technologies is at least equally important, costly and difficult as the innovation of new technologies (see for example Jovanovic 1997). It needs no argument that there will be very limited – if any – impact of technological change on productivity growth or energy-efficiency improvements due solely to the act of innovating new technologies, without them being actually adopted in economic processes. And of course technological change can only help in achieving policy targets with respect to the stabilization or reduction of greenhouse gas emissions insofar as new energy-saving technologies are embodied in capital goods that are actually adopted and used (for an overview see Jaffe et al. 2002). This is however less trivial than it might seem to be.

It is well known that the widespread adoption of existing energy-saving technologies that are available on the market could enable a significant reduction in energy use, especially in the short and medium run (for example de Beer 1998, IWG 1997). At the same time it is known that diffusion of new energy-saving technologies is a lengthy process, that adoption of these new technologies is costly and that many firms continue to invest in old technologies even when new and cost-effective energy-saving technologies are available. In other words, there exists a considerable gap between the most energy-efficient and cost-effective technologies available at some point in time and those that are actually in use, a phenomenon known as the energy efficiency paradox (Jaffe and Stavins 1994, Jaffe et al. 1999).

Recently economic models of economy–energy–climate interactions have been developed in which technological change is endogenous, emphasizing the role of knowledge accumulation in processes of technological change (see for example Grübler and Messner 1998, Goulder and Schneider 1999, Goulder and Mathai 2000, Nordhaus 2002, van der Zwaan et al. 2002). Despite the recognition of adoption and diffusion processes as the essential driving forces behind knowledge accumulation, until now most of these models are characterized by a macroeconomic perspective on aggregate knowledge creation, with much emphasis on innovation through R&D and learning-by-doing processes, the latter usually being a costless by-product of cumulative investment, installed capacity or abatement. Consequently these models ignore to a large extent the microeconomic characteristics of technology adoption and diffusion, such as the sluggish nature of technology diffusion resulting from all kinds of investment barriers – including the costs inherent to learning processes. Therefore we have chosen to focus this book on processes of adoption and diffusion of energy-saving technologies, with much attention on the interaction between microeconomic investment decisions and macroeconomic patterns of energy saving and productivity improvements. In doing so we build upon recent developments in economic theorizing on technological change that stress the role of knowledge accumulation (learning), knowledge spillovers, uncertainty and path dependency in driving processes of technological progress. Hence it is these issues that we will put forward as an important part of the answer to the question why diffusion of energy-saving technologies is often so costly, difficult and therefore slow.

1.4 SLOW DIFFUSION OF ENERGY-SAVING TECHNOLOGIES

Like any paradox, the energy-efficiency paradox suggests by definition a contradiction that seems to exist, and that might disappear if one takes a closer look. This book offers a closer look. In particular we provide a new explanation for the observed slow diffusion of energy-saving technologies and we further explore two explanations that have already been offered in the literature. In short, we suggest a new explanation by emphasizing the role of complementarities between technologies and we further explore the role of learning processes and uncertainty in determining investment decisions and, hence, diffusion patterns. Below, we briefly introduce the role of these different factors in explaining slow technology diffusion.

The Role of Complementarities

With technological complementarity we refer to the existence of imperfect substitutability between technologies. The idea that complementarities between technologies can explain the observed slow diffusion of cost-effective energy-saving technologies is based on the view that 'technological change is not a once-and-for-all-event but a continual process in which many different techniques coexist' (Chari and Hopenhayn 1991: 1144). Many new technologies pass through a life cycle in which they initially complement older technologies, and only subsequently (and often slowly) are substituted for older technologies. Not only at the macro level, but also at the sector level or even at the firm level, the picture of technological change is often one of continuous investment in both old and new technologies, depending on the technology and the type of production process (see for example Mokyr 1990; Rosenberg, 1976, 1982; Young, 1993b). Therefore we will argue in this book that the coexistence of multiple technologies is not only a by-product of past investment decisions – as is the case in virtually all vintage models – but also an essential ingredient of the process of technological change and a matter of deliberate choice. Since technologies differ often not only in terms of their energy productivity but also with respect to other qualities, there are returns to using a mix of technologies. In Chapter 3 we will develop a model in which we show that this 'taste for diversity' causes diffusion to slow down, because it provides an incentive to invest in older vintages, even if new technologies are available that are 'better' when considered in isolation.

The Role of Learning

Studying the interaction between technology diffusion and knowledge accumulation implies a crucial role for learning processes (see for example David 1975). We follow Rosenberg (1982: 121–2) in distinguishing three basic types of learning: learning in R&D stages, learning at the manufacturing stage and learning as a result of use of the technology. Throughout this book we refer to the first type as learning-by-searching, to the second type as learning-by-doing and to the third type as learning-by-using. In addition we distinguish information gathering as a fourth type of learning. Learning-by-searching refers to researchers engaged in (institutionalized) research and development processes, who learn over time to increase the efficiency of their search for innovations. Learning-by-doing refers to producers of a new technology, who learn over time to produce the technology at lower costs and/or to improve the quality of the technology. Learning-by-using refers to users of a new technology, who learn over time to increase their productivity as they gain expertise in how better to use the technology. Information gathering refers to potential adopters who learn about the arrival of a new technology on the market and shape their expectations by collecting information about its performance and its value for their own business. This information is obtained by, for example, observing others who have already adopted the technology and talking to technology suppliers and other experts.

These different types of learning are expected to have different effects on the speed of diffusion of new technologies. Concerning information gathering, the speed of diffusion of course depends positively on this type of learning by potential adopters about new technologies. However in principle the speed of diffusion depends negatively on learning-by-using. Since users of a technology explore the productivity potential of a technology by using it, they build up expertise which will get temporarily lost if they switch to a new technology, thereby creating a barrier for adoption of new technologies. The effect of learning-by-doing on the speed of diffusion is likely to be ambiguous. On the one hand, learning-by-doing leads to the emergence of improved versions of a new technology on the market, inducing adoption because of the improved quality of the technology. However learning-by-doing also creates an 'option value' for potential adopters to postpone investment if they expect new improved versions of the technology to become available in the near future, in order to gain this quality improvement. These different types of learning are included in the different models developed in this book, except for learning-by-searching, because it deals with the nature of innovation processes while we focus on diffusion rather than innovation.

The Role of Uncertainty

Technological change is intrinsically uncertain. When do new technologies arrive? What is their performance? Are they complements or substitutes for existing technologies? The very existence of uncertainty provides a barrier for technology adoption and this barrier becomes higher if investment decisions are at least partly irreversible. The latter suggests that firms can be 'prisoners' of past investments, which in combination with the existence of uncertainty provides an incentive to postpone investment in order to limit the likelihood of regret. In Chapter 4 we will develop a model in which we analyse the effect of this incentive on aggregate energy savings and the effectiveness of investment subsidies, in the presence of a learning process. Traditional investment theories (for example Stoneman and David 1984) suggest the presence of a 'double dividend' associated with subsidizing the adoption of energy-saving technologies that are subject to learning effects. Not only do subsidies induce immediate adoption to meet politically imposed targets, they also induce further technological progress since technology adoption induces the 'take-off' of learning effects. We will show that investment subsidies may give rise to a trade-off between early adoption of relatively inferior technologies on the one hand and late adoption of relatively superior technologies on the other hand. As a result, investment subsidies may be counter-effective in the long run since they might contribute to a lock-in into relatively inferior technologies.

1.5 INTERNATIONAL COMPARISONS OF SECTORAL PRODUCTIVITY PERFORMANCE

In their quest for understanding economic growth, economists have studied the role of accumulation, productivity and technology and their relation to one another. Broad consensus exists that long-run economic growth is caused by technology-driven total factor productivity growth, ever since the uncovering of the Solow-residual pointed to total factor productivity growth or technological change as the main driving force behind economic growth. Concerning the empirical work on the sources of economic growth, this led to the emergence of a growth-accounting tradition, measuring the contribution of various determinants of output and productivity growth (see for example Denison 1967; Maddison 1991). The empirical quest to understand productivity growth has long focused almost exclusively on labour and total factor productivity growth. However, during the last decades energy has increasingly been recognized as a crucial production factor fuelling economic growth, leading to increasing attention being paid to energy productivity developments and to the disentanglement of its determinants. In Chapter 5 we combine insights from both strands of the literature in a simultaneous exploration of energy and labour productivity developments.

A principal aim of this empirical analysis is to trace back macroeconomic developments to developments at the level of individual sectors. The reason for this lies in the fact that understanding trends and cross-country differences in technology-driven productivity performance requires the assessment of productivity performance in individual sectors, since aggregate productivity developments are also the result of structural changes. For this purpose we constructed a new database that merges energy data from the IEA Energy Balances and economic data from the OECD International Sectoral Database (ISDB) and the OECD Structural Analysis Database (STAN) at a detailed sectoral level. The new database covers the period 1970–97 and distinguishes 14 OECD countries and 13 sectors, including ten manufacturing sectors, services, transport and agriculture. Using this database we will document some stylized facts on energy and labour productivity growth in terms of both levels and growth rates across countries and across sectors. Moreover we will decompose aggregate productivity growth performance into a part that is due to shifts in the underlying sector structure and a part that is due to remaining productivity improvements. Finally, we will assess the relation between energy and labour productivity growth rates at different levels of aggregation.

Of course economies not only differ, they also interact. Hence productivity developments are not only determined by developments within a particular country or sector but also by what is happening in other countries and sectors. This raises the question of whether cross-country productivity differences are persistent or whether they will converge to similar levels, for example due to capital accumulation or technology transfers. The concept of productivity convergence has its roots in traditional neoclassical growth theory, with its central notion of a transitional growth path to a steady-state income. The Solow–Swan neoclassical growth model postulates convergence of per capita income, building upon the assumption of diminishing returns to capital accumulation at the economy-wide level. New or endogenous growth theory however yields a more diverse picture concerning patterns of convergence. In this view, cross-country convergence depends on the degree of international knowledge spillovers, allowing less productive countries to catch up with more advanced economies. At the same time, endogenous growth theory suggests that growth differentials may persist or even increase, since learning effects, externalities and market imperfections allow for economy-wide increasing returns to capital accumulation and the existence of multiple steady states. Consequently, the convergence hypothesis has become the subject of extensive empirical research and debate, concentrating on the question of whether initially poor countries indeed grow faster than rich ones. In this book we step back from this debate, but add to the existing empirical macroeconomic literature a systematic comparison of energy and labour productivity convergence at a detailed sector level.

1.6 OUTLINE OF THE BOOK

This book consists of four parts. Part I provides the motivation for the research carried out in this book and provides a short review of economic thinking about economic growth and technological change. Apart from the current chapter it also includes Chapter 2, which compares and contrasts neoclassical and evolutionary thinking. In Part II, containing Chapters 3 and 4, we develop two theoretical models of energy-saving technology diffusion, and the way in which microeconomic characteristics of investment decision determine the impact of environmental policies. Part III, containing Chapters 5 and 6, gives an empirical analysis of energy and labour productivity performance across countries and across sectors. In Part IV, containing Chapters 7 and 8, several elements of the previous chapters are integrated into an existing policy model, conclusions are drawn and some suggestions are made for policy-making and future research. The different chapters are introduced in more detail below.

Chapter 2 provides an interpretative survey of insights from a neoclassical and an evolutionary perspective on economic growth and technological change. We will compare a number of landmark models of both views along three central issues in the debate on technological change, namely the role of heterogeneity, uncertainty and path dependency. What are the essential differences and similarities? What are the recent developments in both traditions and what does this mean for their relation to one another? By answering these questions, this chapter lays the conceptual foundation for the second part of the book.

In Chapter 3 we develop a vintage model that is characterized by 'returns to diversity' of using a mix of technologies. The model is inspired by the product-variety theory which started with the seminal work of Dixit and Stiglitz (1977) and was later extended and applied by for example Ethier (1982), Grossman and Helpman (1991) and Romer (1990). We use the model to offer a new explanation for the observed slow diffusion of cost-effective energy-saving technologies which is rooted in the existence of complementarities between technologies. How exactly will this propensity for technological diversity affect diffusion patterns and adoption behaviour? What happens to this picture if we include a learning-by-using effect? What will be the impact of an energy tax on aggregate energy saving in a world with complementarities and learning-by-using? How does the impact of an energy tax depend on, respectively, the energy intensity of a sector or firm and on substitution possibilities between capital and labour?

In Chapter 4 we develop an investment decision model that acknowledges the importance of uncertainty with respect to both the arrival and the quality of new varieties of a technology in the presence of irreversible investment decisions. The model builds upon the theory of investment under uncertainty (see for example Dixit and Pindyck 1994, Farzin et al. 1998, Pindyck 1991). We include a learning-by-doing effect assuming potential adopters to benefit from the experience of early adopters who provide feedback to the producer, resulting in improved versions of the technology. Hence technological change is endogenous in the sense that the likelihood of a new improved version of the technology arriving on the market is an increasing function of the number of firms that have already adopted the technology. We use the model to explore the impact of uncertainty and learning-by-doing on diffusion patterns and aggregate energy saving, taking into account the existence of an 'option value' for potential adopters to postpone investment. Moreover we examine the effectiveness of investment subsidies in achieving emission reductions under the assumption of uncertain technological change and the irreversible nature of investment decisions. We will show that investment subsidies may be counter-effective in the long run since they can contribute to a lock-in into relatively inferior technologies. We analyse the sensitivity of these results to the nature of technological progress by allowing for different assumptions with respect to learning or spillover effects, the success of innovation, the speed of quality improvement, the discount rate and the existence of physical upper bounds on a technology's energy-saving potential.

Chapter 5 contains an empirical analysis documenting several stylized facts on energy and labour productivity growth in terms of both levels and growth rates. In doing so we explore in particular the role of different sectors in driving aggregate productivity developments. We examine the role of manufacturing, services, transport and agriculture in macroeconomic energy and labour productivity changes, and subsequently explore the manufacturing sector in more detail, examining the productivity performance of ten manufacturing sectors. Are there considerable cross-country and cross-sectoral differences in productivity growth rates? Which part of the aggregate productivity improvement is due to structural changes and which part to technology-driven efficiency improvements? Do patterns of energy productivity growth reflect patterns of labour productivity growth, or do the data suggest that they substitute for each other? What do these results suggest about possible biases in technological change?

In Chapter 6 we examine patterns of cross-country productivity performance and move beyond the descriptive empirical analysis of Chapter 5 by searching for the determinants of sectoral energy and labour productivity developments. We will document the development of cross-country variation of energy and labour productivity over time and we will test for the hypothesis that the productivity of countries with relatively low initial energy and labour productivity levels tends to grow relatively fast, suggesting that being relatively backward in productivity level carries the potential for rapid advance. Are cross-country productivity differentials persistent, or will they converge to each other? And if so, how quickly and by what means? Do energy-inefficient countries catch up with technological leaders? Do advanced economies converge in the long run to a uniform (autonomous) rate of energy efficiency improvement, or does it depend on country- and sector-specific characteristics? Do patterns of energy productivity convergence resemble patterns of labour productivity convergence?

In Chapter 7 we integrate several elements of the previous chapters in the Netherlands Energy Demand Model (NEMO), an existing policy model developed and used by the CPB Netherlands Bureau for Economic Policy Analysis (Koopmans et al. 1999). NEMO is a top-down model describing substitution between capital, labour and energy in the production of energy services, assuming exogenous technological progress, but emphasizing the development of energy efficiency due to adoption and diffusion of energy-saving technologies. NEMO deals explicitly with the energy-efficiency gap by using a bottom-up engineering database to estimate the key parameters of the model, defining the rate of technological progress, substitution possibilities and price elasticities at a sectoral level. We introduce into the model mechanisms representing technological complementarity and learning-by-using processes for new technologies, and examine their effect on technology diffusion and, hence, on aggregate energy use and on the impact of energy policies. Moreover we provide a sensitivity analysis for the potential impact of induced technological change and increasing substitution possibilities between capital, labour and energy.

In Chapter 8 we elaborate upon the main conclusions from this study and provide some suggestions for policy-making and future research.

2. Economic theory of growth and technological change: a neoclassical versus an evolutionary perspective[1]

2.1 INTRODUCTION

Economic theorizing on technological change has its roots in economic theories of economic growth. The traditional, neoclassical model of economic growth was first developed by Solow (1956) and Swan (1956) in the 1950s. In the 1980s, during an economic slowdown, that model was criticized as it explained economic growth, and more specifically technological progress, by simply postulating it. It is therefore not surprising that alternatives to the traditional framework were developed in the early 1980s (see for example Stern 1991). The two major alternatives to the traditional neoclassical growth model are the neoclassical endogenous growth models as pioneered by Romer (1986) and Lucas (1988), and the evolutionary models, for which the foundations have been laid down in the seminal work of Nelson and Winter (1982). Both classes of models are less homogeneous than the 'old' neoclassical model of economic growth. They both strive for a better understanding and a more realistic depiction of the complex process of technological progress. The two classes of models differ with respect to their acceptance of tools and insights from neoclassical economics and their opinion about the extent to which the real world can be described as developing along a smooth path.

In this chapter we discuss the insights and views from both the neoclassical (endogenous growth) perspective and the evolutionary perspective on technological change. In doing so, we adopt Metcalfe's view that 'boundaries are always interesting places to be, but boundary disputes are only occasionally illuminating, and are normally tedious and unproductive' (Metcalfe 1988: 561). Therefore the survey given in this chapter is not so much a dispute but rather a search for the essential differences and similarities between the two points of view. They will be discussed along three central issues in the debate on technological progress:

the importance of, respectively, heterogeneity, uncertainty and path dependency. The organization of this chapter is as follows. Section 2.2 provides an overview of neoclassical models of economic growth and technological change. Both exogenous growth and endogenous growth models are discussed. Section 2.3 presents the evolutionary perspective. Section 2.4 discusses the differences between the two approaches. A final section concludes.

2.2 GROWTH THEORY: A NEOCLASSICAL PERSPECTIVE

This section presents a short overview of economic growth theory from a neoclassical perspective. Rather than providing a detailed survey, its aim is to provide the key characteristics of the different models as well as a brief discussion of the mechanisms that drive growth in both exogenous and endogenous growth models. For more complete surveys of neoclassical growth models we refer to, among others, Lucas (1988), Aghion and Howitt (1998), Barro and Sala-i-Martin (1995), Klundert and Smulders (1992), Nijkamp and Poot (1998), Helpman (1992) and Verspagen (1992).

Exogenous Growth Theory

When thinking about economic growth, a natural starting point is the Solow–Swan growth model (Solow 1956, Swan 1956). This model can be seen as the benchmark for what is now called the neoclassical theory of growth. The Solow–Swan model aims to provide a theoretical framework for understanding worldwide growth of output and the (persistence of) geographical differences in per capita output. Aggregate output (Y) depends on capital (K) and labour (L) according to a constant returns-to-scale production function. Technological progress is introduced in terms of an aggregate parameter (A) reflecting the current state of labour-augmenting technological knowledge. Taking a Cobb-Douglas production function, which is consistent with the stylized fact of constant labour and capital-income shares, we arrive at:

$$Y = (AL)^{1-\alpha} K^{\alpha} \qquad 0 < \alpha < 1 \tag{2.1}$$

The model assumes constant returns to capital (K), effective labour input (AL) and perfect competition. The productivity parameter A is supposed to grow at an exogenously determined constant rate g. The growth rate of the labour force (n), as well as the saving rate (s), are assumed to be constant and exogenous. This latter assumption of an exogenous savings rate is not central to the neoclassical model of growth but simplifies the analysis at this stage; we return to this assumption below. On the basis of these assumptions, an economy, regardless of its starting point, converges to a balanced growth path where long-run growth of output and capital are determined solely by the rate of labour-augmenting technological progress and the rate of population growth. A crucial property of the above production function is that there are diminishing returns to the accumulation of capital at the economy-wide level. The rate of change of the stock of capital per capita (in efficiency units), $k(=K/AL)$, depends positively on the savings rate, and negatively on the rate of population growth, the rate of technological progress, and the depreciation rate (δ), according to:

$$\dot{k} = sk^{\alpha} - (n + g + \delta)k \,. \tag{2.2}$$

A steady state is defined by the situation in which output and capital grow at the same proportional rate as the effective population (AL) resulting in a constant capital intensity (k). This implies that output and capital per capita will grow at the exogenous rate of technological progress (g). In other words, the economy converges to a steady state in which diminishing returns are exactly offset by exogenous technological progress. Modifying the model by allowing for inter-temporally optimizing behaviour of consumers resulting in an endogenous savings rate (Cass–Koopmans–Ramsey), does not alter this basic conclusion (see for example Romer 1986, Blanchard and Fisher 1989).

As already mentioned, both technological progress and population growth are assumed to be exogenous in the Solow–Swan model. Obviously, here we touch upon the main shortcoming of the 'old' neoclassical approach: it takes as given the behaviour of the variables that are identified as the driving forces of growth. That is, it explains growth by simply postulating its existence. Furthermore although the Solow–Swan model fits some stylized facts (cf. Kaldor 1961),[2] when used for growth accounting it turns out that the model is unable to explain growth rates of output by relying on the accumulation of physical inputs (capital and labour); once output growth is corrected for the increase in physical inputs, a large and persistently positive residual remains, the so-called Solow-residual (see for example Maddison, 1991, 1999; Young, 1994, for a recent application of growth accounting to the performance of the East Asian NICs). Therefore factors other than capital accumulation and an increasing labour force should be held responsible for most of the economic growth that has occurred. The Solow-residual is often referred to as the 'measure of our ignorance'. It captures the fundamental driving force behind economic growth, which has been named 'total factor productivity growth' and has often been equated with 'technological progress'. The endogenous growth theory, to which we turn next, aims at contributing to our understanding of the driving forces behind technological progress.

Endogenous Growth Theory

Endogenous growth theory encompasses a class of models that goes beyond Solow–Swan by endogenizing technological change. One of the first attempts to endogenize technology was made by Arrow (1962), who assumed that the growth rate of effective labour input is a result of workers' cumulated experience in producing commodities, or in other words, the result of learning-by-doing. This implies that the labour productivity is now endogenous, being an increasing function of cumulated aggregate investment by firms. An important characteristic of the Arrow model is that learning is conceived as a public good; it is the result of experience at the level of the whole economy and can be applied by all firms at no cost. This also means that in deciding how much to invest, firms ignore the effect of their investment on the total amount of knowledge in the economy since the effects are external to each individual firm.

A major step forward in endogenizing technological progress was made by Romer (1986), who builds upon the contributions of Frankel (1962) and Arrow (1962). The basic idea of his approach is that technology grows in proportion to the macroeconomic capital stock, potentially offsetting the effects of diminishing returns. Capital in such a setting should be considered as a broad concept, including human and intangible capital. This approach is currently known as the 'AK approach' because it results in a production function of the form $Y = AK$ with A constant. The individual firm's i production function reads:

$$Y_i = [A_i(K,L)L_i]^{1-\alpha} K_i^{\alpha} \tag{2.3}$$

Crucial in this formulation is that the index of knowledge available to the firm is linked to the economy-wide stock of capital and labour. This is based on the idea that each firm's knowledge is a public good that any other firm can access at zero costs. Accordingly, the change in each firm's technology term A_i is related to the change in the aggregate capital and labour stock, which the individual firm cannot influence (they are external to the firm). A simplifying specification for A_i that is often used is:[3]

$$A_i = [AK^{\beta}]^{\frac{1}{1-\alpha}} \quad (2.4)$$

in which the change in technology only depends on the rate of macro-economic capital accumulation. Aggregating over all firms, we arrive at the aggregate production function:

$$Y_i = AK^{\alpha+\beta} L^{1-\alpha} \quad (2.5)$$

In the special case in which $\alpha+\beta=1$ and L is normalized at one, the production function boils down to $Y=AK$. Long-run economic growth can be sustained without relying on exogenous technological progress.[4] The rationale for this approach, in which technological development or learning is external to the firm, lies in the difficulty of dealing with increasing returns in a general equilibrium framework. By introducing a Marshallian externality, a competitive equilibrium, in which capital and labour receive their marginal products, can exist. In other words there exist constant returns to scale at the firm level and increasing returns at the economy level due to increasing knowledge.

In conclusion, the essential idea of the Romer (1986) model is that knowledge can be considered as a kind of renewable capital good, where K should be interpreted as knowledge. Within the model, long-run growth is determined by the (still unintended) accumulation of knowledge through representative individuals who maximize inter-temporal utility (Ramsey model). The crucial assumption in this Romer model is that knowledge does have a non-decreasing marginal product, that is $\alpha+\beta \geq 1$. This can be interpreted as allowing for non-decreasing social returns to capital (knowledge) resulting in non-decelerating growth. In contrast with the Solow–Swan model of growth, a positive growth rate of output can be sustained without population growth and exogenous growth of technological progress.

It should be noted that this class of models has a knife-edge character due to the assumption of constant returns to scale with respect to reproducible factors. Any deviation from this assumption will have significant effects in the (very) long run. With slightly decreasing returns, growth will vanish in the very long run, whereas with slightly increasing returns growth will accelerate indefinitely (as in Romer 1986). As a consequence, the explanatory power of endogenous growth models with respect to changing technology variables for the very long run cannot be taken too literally. One cannot preclude that reality is characterized by slightly increasing or decreasing returns to scale.[5] Nevertheless it can be argued that this weakness is of relatively minor importance, since with sufficiently large elasticities of reproducible factors (which seem to be the empirically relevant case), transition periods in which technology variables change are very long (see also Mankiw et al. 1992 for a discussion of the speed of transition in the context of the neoclassical model of growth). This implies that endogenous growth models, despite their knife-edge character, yield useful insights into the process of endogenous technological change over long periods of time, although not indefinitely. We refer here also to McGrattan (1998) for a recent convincing theoretical and empirical defence of the AK growth models.

As a way of summarizing, Figures 2.1a and 2.1b present the steady-state outcomes of, respectively, exogenous and endogenous models of economic growth with Ramsey consumer behaviour.

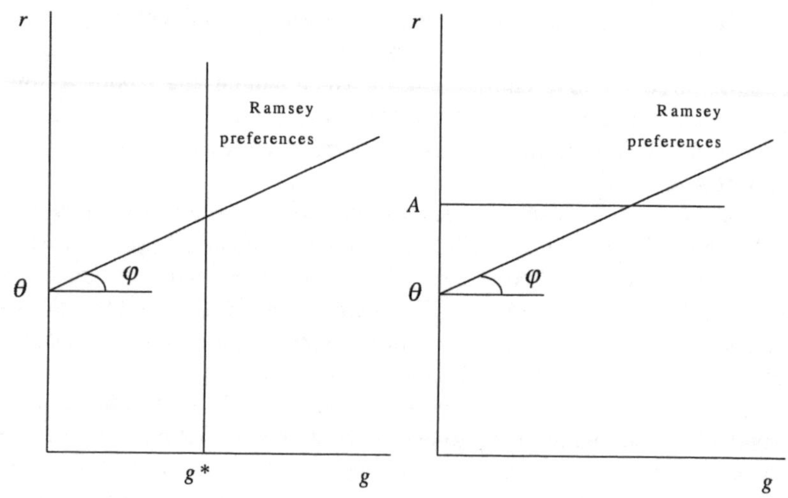

Figure 2.1a *Figure 2.1b*

Figure 2.1 Steady-state outcomes of the neoclassical model of exogenous growth (2.1a) and the AK model of endogenous growth (2.1b)

Figures 2.1a and 2.1b basically depict equilibrium on the capital market, where funds are supplied by consumers who save part of their income with the aim of smoothing their consumption profile over time. Funds are demanded by producers who need money for investment in growth-promoting activities. The supply of funds by consumers satisfies the Ramsey rule according to which there is a positive relationship between the growth and the interest rate. The positive slope is due to the fact that consumers who wish to smooth their consumption profile are only willing to accept a high growth rate (that is, a steep consumption profile) provided that they are compensated by a high interest rate. This relationship holds independently of whether we have an exogenous or an endogenous growth model. Assuming a constant pure rate of time preference (θ) and coefficient of risk aversion (φ), one arrives at the Ramsey formula $g = (r-\theta)/\varphi$. On the demand side of the capital market are the producers who need funds to finance their investments. In the neoclassical (exogenous) growth model, the demand for funds is represented by a vertical line at the growth rate equal to the exogenous rate of technological progress (g^*). This demand for funds is independent of the interest rate r (Figure 2.1a). The intersection of this line with the line representing the Ramsey rule yields the solution for the steady-state interest rate (that is the steady-state marginal product of capital). In the basic AK approach as set out above, the long-run growth rate of capital and output are determined endogenously, whereby the demand for funds satisfies the condition that the marginal product of capital equals A (that is, the relationship between the growth and the interest rate is horizontal in Figure 2.1b). Again, the equilibrium growth rate is found at the intersection of this line with the line representing the Ramsey rule (Figure 2.1b).

Although the presence of externalities and non-diminishing returns to scale in the final goods sector is an important characteristic of the basic AK model of endogenous growth, subsequent developments have generalized the basic model in several ways. Rebelo (1991) demonstrated the feasibility of sustained endogenous growth without the need for constant returns to scale with respect to reproducible factors in the final goods sector, and without the presence of externalities. The underlying idea is that all that is necessary is a 'core' of reproducible capital goods or knowledge entering into final goods production (with potentially diminishing returns) that is produced in some sector of the economy without the direct or indirect contribution of non-reproducible factors and with (at least) constant returns to scale.

Yet another route that has been taken in the development of endogenous growth models is the modelling of growth as resulting from intentional microeconomic activities aimed at profit maximization in the context of imperfectly competitive general equilibrium models. These models no longer rely on pure externalities resulting (unintentionally) in non-diminishing returns with respect to accumulative factors at a macroeconomic level (for example Lucas 1988, Romer 1987). An important issue in these models is the extent to which the returns to investment are appropriative and the extent to which they spill over to other agents in the economy.

A third route that has been explored is the development of multi-sector models. These models are obviously not only more realistic, but they are also of importance when non-reproducible inputs are essential in the production process. Technological progress is then needed to counterbalance the tendency of ceasing long-run growth through decreasing returns to those factor inputs. The usual way to model intentional technological progress is to make a distinction between a research sector and other sectors. The allocation of resources between the R&D sector and the other sectors is determined by relative returns (for example Grossman and Helpman 1991). The important characteristic of all models of endogenous growth is that the rate of growth can, in contrast with the traditional neoclassical model, be influenced by economic policies (even in the long run). Many of the models also require government intervention from a socially optimal point of view due to the existence of externalities such as knowledge spillovers (intertemporal, as well as between firms at one point in time) that are characteristic for many endogenous growth models.

The most recent development in endogenous growth theory is the so-called neo-Schumpeterian approach to endogenous growth theory. The essential characteristic of this approach is the existence of 'creative destruction', that is the occurrence of a succession of innovations in one or more sectors, resulting from research activities, and implying a business stealing effect. An example of a Schumpeterian model of endogenous growth is in Aghion and Howitt (1992), where growth is fuelled by a random sequence of quality-improving innovations. The innovations 'arrive' according to a stochastic Poisson process and result from uncertain research activities. One of the implications of this model is again that the laissez-faire growth rate may be different from the socially optimal growth rate. The reason for this is the existence of three counteracting effects: the business stealing effect, the intertemporal spillover effect and the appropriability effect. The latter refers to the private monopolist's inability to appropriate the whole output flow. Both the appropriability effect and the intertemporal spillover effect tend to generate insufficient research under laissez-faire relative to what is socially optimal because private research 'firms' do not internalize positive social externalities. The business stealing effect on the other hand tends to generate too much research under laissez-faire relative to what is socially optimal because it ignores the costs of obsolescence of old intermediate inputs. In the case that the business stealing effect dominates, laissez-faire growth will be excessive. Again this implies a rationale for government intervention.

2.3 GROWTH THEORY: AN EVOLUTIONARY PERSPECTIVE

Dissatisfaction with the Solow–Swan model of economic growth (see section 2.1) has not only led to the wave of endogenous growth models as described in the previous section, but also to evolutionary theorizing on economic growth. The benchmark study here is the classic book of Nelson and Winter (1982), which exposes an evolutionary view on economic growth and technological change. The work of Nelson and Winter can be seen as an 'evolutionary response' to the shortcomings of the old neoclassical theory of economic growth. The development of evolutionary growth models resulted in a revival of the evolutionary tradition in economics, which goes back to Veblen (1898) and Schumpeter (1934). The idea that technological change is a fundamental driving force of economic development is at the heart of evolutionary theorizing about economics in general and economic growth in particular. Of course the crucial role of technological progress in bringing about economic growth has been recognised in neoclassical economics as well. However what distinguishes evolutionary from neoclassical economics is its theoretical framework.

The differences between the two theoretical frameworks arise essentially from the objections evolutionary economists have to the (aggregate) production function as it is used by neoclassical economists. The neoclassical production function can be defined as a specification of all conceivable modes of production in light of the existing technological knowledge about input–output relationships (Sahal 1981). It is common practice among neoclassical economists to distinguish between a movement along the production function, referring to factor input substitution, and a shift of the production function, referring to technical progress. Among others, Nelson and Winter (1974, 1977, 1982), Dosi (1982) and Sahal (1981) have argued that this view of production and technological progress suffers from providing no insight into the occurrence of technical innovation processes, because the development of new techniques ('blueprints') is exogenous to the economic process (a criticism that does not apply to the class of endogenous growth models). Furthermore evolutionary economists argue that the neoclassical production function does not comply with what empirical research tells us about the nature of technological change and the characteristics of innovative firms.[6]

Following Dosi (1988), the major findings of the (microeconomic) literature on sources, procedures, directions and effects of technological change can be summarized as follows:

1. Innovative efforts are characterized by varying degrees of appropriability and uncertainty about the technical and commercial outcomes.
2. Technology embodies a certain degree of tacit knowledge that is firm-specific, local and cumulative ('expertise').
3. Innovations result from search and learning processes of individuals or organizations (firms). Knowledge-building and problem-solving activities are characterized by organizational or behavioural routines ('bounded rationality'; 'satisficing').
4. As a result of (2) and (3), technologies develop along relatively ordered paths ('trajectories') within the boundaries of firms or technological paradigms. The latter refers to a collective framework that determines the R&D practice and the pattern of technological development on the basis of a dominant design of an artefact. Technologies may well get locked in to paradigms, implying a certain degree of irreversibility ('history matters').
5. As a result of (1), (2), (3) and (4), the diversity of techniques used within and between firms and sectors is a fundamental characteristic of an economy undergoing technological change.

On the basis of these findings a whole body of literature has been developed within evolutionary economics addressing the 'evolution of the production function'. In the literature the occurrence of technical innovation is made endogenous according to an evolutionary line of reasoning. An in-depth discussion of the literature embodying evolutionary theorizing on technical change is beyond the scope of this book. For lucid surveys we refer to Cimoli and Dosi (1995), Dosi et al. (1988), Dosi and Nelson (1994), Nelson (1995), Silverberg (1997) and Nelson and Winter (2002).[7]

With respect to evolutionary theorizing on economic growth, the basic formalization draws on what are now called 'replicator equations' (Silverberg and Verspagen 1998). In biology these equations were introduced by Fisher (1930) and can be seen as a formalization of Spencer's notion of the survival of the fittest.[8] Replicator equations describe the evolution of a population according to:

$$\dot{x}_i = x_i(f_i(x) - \bar{f}(x)), \quad i = 1,\ldots,n \tag{2.6}$$

$$\text{with } \bar{f}(x) = \sum_{i=1}^{n} x_i f_i(x) ,$$

where x is the vector of relative frequencies of the species $(x_1,\ldots,x_n)$, n reflects the number of distinct competing 'species' of a population and $f_i(x)$

reflects the frequency-dependent fitness of species *i*. This implies that species with above-average fitness will expand in relative importance, those with below-average fitness will decrease, while the average fitness $\bar{f}(x)$ changes with the relative population weights. Assuming constant fitness functions, it can be shown that the species with the highest fitness will displace all the others and that average fitness is dynamically maximized by the evolutionary process.[9] When fitness functions change endogenously, the replicator equation describes relative share dynamics. In evolutionary economic models, $f_i(x)$ is generally defined in terms of market competition or differential profit rates, thereby functioning as the selection mechanism. The entities (*i*), on which selection takes place, are 'translated' to firms, techniques or products (Payson 1995), while the share of competing 'species' of a population level x_i usually refers to the market shares.

As a result, evolutionary economic models are characterized by an explicit micro-foundation, allowing for heterogeneity. This is an important characteristic that also holds for evolutionary growth models. In the Nelson and Winter (1982) growth model, heterogeneity is defined in terms of firms: different firms possess different capabilities, procedures and decision rules that determine their action. Firms use production techniques (characterized by fixed labour and capital coefficients) to produce a homogeneous output. Novelty, here defined as process innovation, is a result of search activities of individual firms, with 'search' denoting a firm's R&D activities. The search process ('mutation') may consist of imitation or local search. In the first case, firms explore techniques already used by other firms. In the second case, firms search for new techniques in a given and finite pool of existing but yet undiscovered techniques. The search process is guided by satisficing behaviour (Simon 1955): firms only undertake a search process if their rate of return on their employed techniques falls below an arbitrarily determined value. The rate of return on techniques drives the selection process in the following way. Assuming that techniques differ only with respect to their labour productivity and that firms reinvest their profits in capacity expansion, the more productive and profitable technique gradually replaces the less productive one, depending on the form of the wage function.

The Nelson and Winter model is analysed by means of simulation because of the complexity arising from the simultaneous existence of multiple firms with different search behaviour, and hence different technological levels. Nelson and Winter show that their model is capable of generating the patterns of aggregate output (GNP), capital input, labour input and factor prices, as addressed by Solow (1957) in his contribution to growth theory. Although the empirical validation of the Nelson and Winter model is highly specific to Solow's data set, Nelson and Winter have argued on the basis of this outcome that an evolutionary model is preferable to a neoclassical one since it may explain macroeconomic facts about productivity growth, 'through the route of "building them up" from micro-economic data' (1982: 209). It is mainly this argument that has given the Nelson and Winter model its paradigmatic function of being the starting point for a revival of evolutionary economic theorizing.

A number of evolutionary growth models have been developed that are inspired by the Nelson and Winter model. The most important common factor in these models is the modelling of a selection process, on the basis of replicator equations, as the driving force of economic growth. The selection process continually modifies the technological variety, which may be embodied in firms, sectors or countries.

Conlisk (1989) has developed a simple aggregate growth model in which technological change is endogenous and 'cumulative'. The latter refers to the Nelson and Winter model and the replicator dynamics in the following way. Technological change is defined in terms of the productivity of firms. The economy constructs new firms with productivities that distribute around the mean of a normal probability distribution. A new firm with a productivity that is sufficiently higher than the average productivity is viewed as an innovation that improves standards for future firms. This implies that technological change increases the mean of the productivity distribution.[10] Since the new firm's technology is assumed to build on the innovative firms of the past, rather than on the average firms of the past, innovation causes productivity to grow at the economy level. The growth rate of the economy is a function of the savings rate, the speed of diffusion of new knowledge and the standard error of the productivity distribution of new firms. The latter can be interpreted as the average innovation size. Due to a number of simplifications, this model is one of the very few evolutionary growth models that yields analytical solutions.

In Silverberg and Lehnert (1993, 1994), technological variety arises according to a time-homogeneous or inhomogeneous Poisson distribution. Again the replicator mechanism is driven by profit rates. As in the Nelson and Winter model, techniques with different labour productivity yield different profit rates, and profits are subsequently reinvested in those same techniques. Hence techniques with above-average productivity will expand, new techniques diffuse into the economy due to their superior profitability and techniques with below-average productivity (and hence profitability) will contract. This led the authors to label their model 'Schumpeterian', referring to Schumpeter's notion of creative destruction. More specifically, the stock of capital (k) embodied in a qualitatively distinct technology (i) evolves according to:

$$\frac{\dot{k}_i}{k_i} = r_i + s(r_i - \bar{r}) - \gamma = \frac{1}{c_i}\left(1 - \frac{w}{a_i}\right) + s(r_i - \bar{r}) - \gamma \qquad (2.7)$$

where r_i is rate of profit of technology i, $\bar{r}$ the capital stock weighted average profit rate, s a constant representing the strength of the cross-technology investment, γ the depreciation rate of physical capital, c_i the capital output ratio, a_i the labour productivity and w the real wages. Essentially the model is a multi-technique version of Goodwin's growth cycle model (Goodwin 1967) and therefore mathematically equivalent to a multi-dimensional Lotka–Volterra system with stochastically perturbed coefficients.[11] In other words, the mathematical method on predator–prey relationships is used to derive the selection process. Let f_i be the share of technology i in total capital stock, where technologies with above-average profit rates grow in share according to:

$$\dot{f}_i = (1+s)(r_i - r)\,. \qquad (2.8)$$

As a result long-run fluctuations in productivity growth are produced within the model, basically driven by innovations. In this context a reference is made to earlier work of Schumpeter and Kondratieff on the relationship between innovation, diffusion and macroeconomic dynamics.[12]

In the growth model of Verspagen (1993), heterogeneity is defined in terms of sectors. Sectors differ with respect to produced goods and labour productivity. Again the latter is an indicator for the technology used. A replicator equation in which sectors compete with each other on the basis of production costs (consisting of technological, labour and exchange rate costs) is responsible for the selection process. This multi-sectoral model highlights the interaction between specialization and growth by making technological performances of countries dependent on their specialization pattern. Simulation outcomes show that differences between technological potentials of sectors and countries may give rise to persistent differences in growth rates.

Following the Nelson and Winter approach, a number of models have been developed that take an explicit micro-economic foundation, that is, in the meaning of modelling individual firm behaviour (for example Chiaramonte and Dosi 1993, Dosi et al. 1994, Silverberg and Verspagen 1994a, b). In these models differences between firms arise not only from technological differences (in terms of input coefficients), but also from economic strategies followed by firms. The latter refers to pricing of products (based on demand expectations) or decisions about the investment of resources in the search for new techniques. In general the probability of innovation depends on R&D efforts. These models differ from one another with respect to, among others, whether or not firms possess skill levels that evolve by a learning process, R&D strategies that can change over time, and R&D or technological spillovers that play a role.

2.4 NEOCLASSICAL VERSUS EVOLUTIONARY PERSPECTIVES: SEARCHING FOR THE ESSENTIAL DIFFERENCES

As noted above, recent evolutionary theorizing on economic growth and technical change evolved out of dissatisfaction with the empirical performance of the traditional neoclassical approach to economic growth. In the words of Nelson and Winter: 'The weakness of the (neoclassical) theoretical structure is that it provides a grossly inadequate vehicle for analysing technical change. In particular, the orthodox formulation offers no possibility of reconciling analyses of growth undertaken at the level of the economy or the sector with what is known about the processes of technical change at the microlevel' (Nelson and Winter 1982: 206). Obviously part of the criticism of neoclassical models of economic growth has been superseded by the development of the endogenous growth theories that are in accordance with the neoclassical framework. Still, from an evolutionary point of view a number of objections or criticisms to the new neoclassical theories persist (Nelson 1994). Essentially these can broadly be summarised in two points:

1. A large gap still exists between the stylized and formal treatment of technologies in the endogenous growth models based on rational and optimizing behaviour and what is known from microeconomic studies on technology and technological change, firm behaviour and institutions. In other words, new neoclassical growth theory still contains an overly 'stylised, stripped down, inadequate view' (Nelson 1994: 319).
2. The new neoclassical growth models continue to treat economic growth as a smooth process involving a continuous tendency to return to an equilibrium state.

In this section we explore the differences between the evolutionary and the neoclassical approach to economic growth and technological development. The aim of this section is to investigate to what extent the differences are essentially in terms of modelling. We have chosen to discuss the differences between the two approaches along a classification of issues that are central to the debate on technological progress, namely the view on technology, heterogeneity, uncertainty and path dependency. By taking these issues into account explicitly, evolutionary economics claims to distinguish itself from neoclassical economics (Nelson and Winter 1982). For this reason we will investigate in particular the value added of evolutionary growth models in terms of their outcomes on the basis of these issues.

Technology

In the Solow–Swan model technology is incorporated as an exogenous factor representing knowledge as a non-rival or public good. In the initial attempts to endogenize technology in growth models, several definitions of knowledge have been given. In the Arrow model, knowledge creation is defined as 'learning-by-doing'. Although still a public good for individual agents, knowledge is supposed to increase as an unintended side-effect of production activities at the economy level. In other neoclassical contributions, knowledge is interpreted as scientific knowledge, characterized by lack of both private returns and property rights (Phelps 1966; Nordhaus 1967; Shell 1966, 1967). It is this view of technology that has been heavily criticized by evolutionary economists.

As we stressed in section 2.2, an essential characteristic of the wave of endogenous growth models is a broadening of the capital definition. This is mainly done by defining capital not only in terms of physical capital, but also in terms of human capital. Knowledge accumulation, in one way or another, is at the heart of all models on endogenous economic growth, since it is responsible for sustained endogenous growth. For example, in Arrow (1962), Romer (1986), Lucas (1988) and Rebelo (1991), knowledge is defined as the factor responsible for increasing capital productivity, with capital broadly defined as encompassing both physical and human capital.

Although the early approaches in endogenous growth theory modelled technological progress as a side-effect of economic activities, recent endogenous growth models explicitly take into account intentional technological progress. For example in Grossman and Helpman (1990, 1991) and Smulders and van de Klundert (1995) knowledge results from intentional R&D activities. The return on investment in research activities consists of monopoly rents generated by blueprints of new intermediate inputs that result from R&D activities. Allowance is made for innovative efforts that are characterized by varying degrees of appropriability and technology that may have a certain degree of tacitness (Smulders and Van de Klundert 1995, Peretto 1996). In the Schumpeterian approach (Aghion and Howitt 1992, 1998) a more vertical way of modelling technological progress is employed, taking into account the obsolescence of old intermediate outputs as an integral component of technological progress. Allowance is also made for uncertainty in the arrival of new technological opportunities.

Following Nelson and Winter (1982), evolutionary growth models also embody a Schumpeterian perspective on economic growth and technological progress. Whereas the 'neoclassical Schumpeterian' type of model is still an equilibrium model including typical neoclassical decision-making principles such as profit maximization and rational expectations, the 'evolutionary Schumpeterian' type of models drop these features. Further differences between the neoclassical and evolutionary models of growth fuelled by innovation and 'creative destruction' can be traced back to the way in which heterogeneity, uncertainty and path dependency are considered. In evolutionary growth models these issues are presented as crucial components of economic growth and technological progress. Evidently this results in a more diverse and complex representation of the Schumpeterian view than in endogenous growth models (see for example Chiaramonte and Dosi 1993, Silverberg et al. 1988). Below we will investigate to what extent the evolutionary contribution to growth theory has gained additional insights by elaborating on these issues. Still, it is fair to conclude that over the last few years the views on technology expressed by evolutionary and neoclassical economists have converged.

Heterogeneity

The modelling of differences between technologies, sectors or countries can be considered as an important feature of evolutionary growth models. By modelling this heterogeneity, evolutionary economics claims to provide a more sophisticated microeconomic foundation for explaining macroeconomic patterns. In general microeconomic diversity is defined in terms of technological differences that influence macroeconomic patterns by means of the selection process, as described in section 2.3.

In Conlisk's model (Conlisk 1989), heterogeneity plays a crucial role. Heterogeneity is defined as the degree to which firms differ with respect to their technological features, measured in terms of labour productivity. As described in section 2.3, innovations increase the average productivity of the firms, which is modelled as the mean of the probability distribution reflecting the productivities of the firms. The specification of this mean is critical to the outcome of the model.

A more sophisticated representation of heterogeneity can be found in the models that take into account behavioural differences among firms. For example Chiaromonte and Dosi (1993) model a complex disequilibrium process of innovation, imitation and competition, where the aggregate patterns of productivity and growth are interpreted as the outcome of an evolutionary or self-organizational process driven by endogenous learning and market selection. In this model, as well as in Dosi et al. (1994) and Silverberg and Verspagen (1994a, b), behavioural and technological diversity among agents is a crucial ingredient of the system dynamics. The behavioural diversity is modelled by assigning fixed decision rules to agents, which are different in their decision parameters. These decision rules attempt to reflect what is known from the literature on innovation processes, as we have summarized in section 2.3. In Chiaromonte and Dosi (1993), the most concrete effect of modelling heterogeneity among the agents on aggregate dynamics is the instability of the relationship between productivity growth and income growth that appears from simulations. By way of explanation, it is suggested that variations in appropriability, above a certain minimum threshold, have ambiguous effects on growth. Although the model simulation shows that appropriability is a necessary condition for innovation, since it is obviously beneficial to individual innovators, appropriability may also slow down technological diffusion and thereby aggregate growth in productivity and income.

In Verspagen's model (Verspagen 1993), heterogeneity is defined in terms of sectors that differ with regard to the produced product and labour productivity (technology). At the aggregate level, selection is a function of sectoral shares in total consumption, which evolve according to different real-income elasticities in different countries. In Chiaromonte and Dosi (1993) two related sectors are modelled. One sector produces heterogeneous production inputs ('machines') with labour, the other sector produces a homogenous consumption good with labour and 'machines'. In both models sectoral heterogeneity enables an interaction between the supply and demand side of the economy. Innovation and diffusion processes in the sector that produces input goods are driven by the demand of the other sector, which produces consumption goods.

It is to be noted that the 'neoclassical Schumpeterian' literature also addresses the ambiguous relation between appropriability and productivity growth. In Aghion and Howitt (1998, Chapter 7), the issue is discussed in terms of the relationship between competition and growth. A distinction is made between competition in the product market and in the innovation sector. The analysis shows that competition in the innovation sector is likely to correlate positively with growth, whereas product market competition features an ambiguous relationship with productivity growth. It is suggested that for a better understanding of the trade-off between the positive and negative appropriability effects, it is necessary to classify the economy into subgroups. In short, both approaches to Schumpeterian dynamics stress the importance of heterogeneity, but so far this feature has only been elaborated upon in 'evolutionary Schumpeterian' growth models.

In summary, it appears that evolutionary growth models deal with technological and behavioural diversity in a more sophisticated and explicit way than neoclassical growth models. In the latter class of models, the emphasis is on one- or two-sector economies with representative agents and commodities. The simultaneous consideration of firms or sectors with different technological characteristics gives the evolutionary approach its added value here. However it has to be noted that this level of microeconomic diversity comes at a certain price. Due to the complexity of the models, which precludes analytical tractability, the mechanisms behind the aggregate dynamics are not always clearly exposed (see for example Chiaromonte and Dosi 1993).

Uncertainty

Evolutionary growth models are non-deterministic models, representing the uncertainty inherent in evolution in general and to innovation (diffusion) processes in particular. Usually the models include a certain degree of uncertainty by representing the innovation process as a stochastic process. In the Nelson and Winter (1982) model the probability of discovering techniques is represented as a Markov process. In Conlisk (1989), where innovation is defined in terms of newly entering firms, innovations are generated as random drawings from a normal probability distribution. Motivated by an empirical literature review, Silverberg and Lehnert (1993, 1994) have used three stochastic processes to model the occurrence of innovations. They model a homogenous and a logistically Poisson process as well as an innovation arrival rate that depends on economic variables (lagged profits). Chiaromonte and Dosi (1993) model innovation as a two-stage stochastic process. The first stage refers to the access to the creation of a new technological paradigm (a set of technological typologies). The second stage refers to the addition of a new typology to the set, once access to the paradigm has been established. In addition, actual access to a machine and imitations of machines of other firms occur in similar two-stage stochastic processes.

In short, uncertainty is at the heart of the 'evolutionary Schumpeterian' growth models since technological progress is essentially modelled as a stochastic process. However it is to be noted that uncertainty may also play a role in a mainly deterministic setting. This has been shown for example in the neoclassical theory where attention has been paid to the Schumpeterian notion of 'creative destruction' (see Aghion and Howitt 1992) and to investment under uncertainty and irreversibility (see Dixit and Pindyck 1994). The former captures innovation as a stochastic process resulting from uncertain research activities. In the theory of investment under uncertainty and irreversibility, uncertainty is included as a condition under which deterministic decision-making takes place. The essential idea is that in the case of irreversible investments, it may be rational not to invest immediately in technologies with an expected positive net present value, but to postpone the investment until new information becomes available. Since additional information reduces the chance that investments will be made in projects that turn out to be unprofitable *ex post*, the net present value of a postponed investment can be higher than the net present value of an earlier investment. In other words the combination of uncertainty with irreversibility will result in situations in which investments are negatively related to the degree of uncertainty.[13]

In conclusion, both the evolutionary and the neoclassical literature address the issue of uncertainty, and convergence has taken place in modelling innovation as a stochastic process. Still, the role of uncertainty has mainly been elaborated in 'evolutionary Schumpeterian' growth models. The other side of this picture is that making stochastic processes the crucial mechanisms that drive innovation and diffusion may prevent the models from illuminating the actual mechanisms determining an aggregate pattern of technological change (see for example Silverberg et al. 1988).

Irreversibility and Path Dependency

The issues of path dependency and lock-in are characteristics of technological innovation and diffusion processes and thus important products of evolutionary economic theorizing. Path dependency refers to the role of historical events in determining the adoption of technologies (see David 1985). This implies a certain degree of irreversibility, and hence the possibility that technologies get locked in to a technological paradigm (Arthur 1989).

In growth theories, these topics have been addressed in basically two ways. First, irreversibility in the 'evolutionary Schumpeterian' growth models is generally implicit to the notion of 'creative destruction', as it is in the notion of the business stealing effect in 'neoclassical Schumpeterian' growth models. That is, innovation and diffusion of technologies displace older technologies, imposing irreversible losses to the agents that are locked in to outdated technologies (see for example Aghion and Howitt 1998, Silverberg and Lehnert 1994). In line with this, the existence of vested interests among agents specialized in the old technologies may cause resistance to the adoption of new technologies, and hence slow down technological progress (see Olson 1982; Mokyr 1990, 1992 for many historical examples; and Canton et al. 2002 for a theoretical approach to this issue). It is also here that 'evolutionary Schumpeterian' and 'neoclassical

Schumpeterian' models come close together since they both stress the role of increasing returns to adoption, returns on learning-by-doing, network externalities and technological spillovers in establishing vested interests and lock-in situations. Second, irreversibility and path dependencies play a role in growth theory in terms of permanent effects of temporary shocks on macro-dynamics.

2.5 CONCLUSIONS

Technological change is the main driving force behind economic growth, which is reflected by a close relationship between economic theorizing on technological change and economic growth. Since the 1980s, a renewed interest in economic growth theory has emerged, that has focused on the economics of technological change. Dissatisfaction with the old neoclassical theories of economic growth, rooted in the contributions of Solow (1956) and Swan (1956), has led to a new class of neoclassical and evolutionary models. The first class of models has been labelled endogenous as their key feature is to make technological change endogenous. Following the work of Nelson and Winter (1982), the second class of models endogenize technological change as well, but according to an evolutionary view of economic growth and technological change, emphasizing the importance of heterogeneity, uncertainty and path dependency.

The two approaches converge in what is called a Schumpeterian view on economic growth and technological change. The essential characteristic of this approach is the occurrence of a succession of innovations in one or more sectors, resulting from research activities, and implying a business stealing effect. Differences between the so-called evolutionary Schumpeterian and the neoclassical Schumpeterian growth models can mainly be traced back to the way they deal with heterogeneity, uncertainty and path dependency. It can be argued that in evolutionary models, technological and behavioural diversity, uncertainty, path dependency and irreversibility are elaborated in a more sophisticated and explicit way than in neoclassical growth models. However it has to be noted that this level of microeconomic diversity comes at a certain price. Due to the complexity of the models, which precludes analytical tractability, the mechanisms behind the aggregate dynamics are not always clearly exposed. In other words, although the evolutionary tradition in economics claims to enter further into the technological 'black box' (Rosenberg 1982), it does not always avoid the pitfall of substituting Solow's 'manna from heaven' for 'parameters from heaven'.

Nelson and Winter have formulated the challenge to an evolutionary formulation as follows: 'it must provide an analysis that at least comes close to matching the power of neoclassical theory to predict and illuminate the macro-economic patterns of growth. And it must provide a significantly stronger vehicle for analysis of the processes involved in technical change, and in particular enable a fruitful integration of understanding of what goes on at the micro level with what goes on at a more aggregated level' (Nelson and Winter 1982: 206). It can be argued that although evolutionary modelling has evolved since the Nelson and Winter model, evolutionary growth theory has demonstrated relatively little progress in terms of formal theorizing. So far evolutionary growth models have not yielded much new insight into, or a better understanding of, the mechanisms driving macroeconomic patterns of economic growth and technological change. A challenging task for further research is to combine the fruitful insights of both approaches to improve our understanding of the complex processes of technological change in relation to other macro- and microeconomic processes.

In the next part of this book we build upon the recent developments in economic theorizing on technological change described in this chapter by developing two small analytical models of technology adoption. The aim of these models is to improve our understanding of slow diffusion of new energy-saving technologies, taking into account the role of knowledge accumulation (learning), knowledge spillovers, uncertainty and irreversibility in determining investment decisions. In Part III we move to empirical analyses, measuring the role of technological change in driving productivity growth. These analyses build upon empirical work on the sources of economic growth, which is inspired by the developments in growth theory as they were described in this chapter. The empirical research includes a central issue in the growth literature, which is the convergence of cross-country differences in productivity performance. Finally, in Part IV we integrate several theoretical insights concerning technological change as they will be developed in Part II, in an applied economic model used for policy analysis.

NOTES

1. This chapter is a slightly revised version of the publication: Mulder, P., H.L.F. de Groot and M.W. Hofkes, 2001, 'Economic Growth and Technological Change. A Comparison of Insights from a Neo-Classical and Evolutionary Perspective', *Technological Forecasting and Social Change*, 68, pp. 110–34. Reprinted with permission from Elsevier.
2. These facts are that (1) the growth rates of labour, capital and output are each roughly constant; (2) the growth rates of capital and output are roughly equal; (3) the growth rates of capital and output are bigger than the growth rate of labour, implying that output per worker and capital per worker are increasing over time.

3. A more general specification would be $A_i = [AK^{\beta}L^{-\varphi}]^{1/(1-\alpha)}$. We then arrive at $Y=AK^{\alpha+\beta}L^{1-\alpha-\varphi}$. So again long-run growth can be sustained with constant population provided that $\alpha+\beta \geq 1$.
4. The analyses of Romer (1986) and Arrow (1962) essentially differ in that Arrow still assumed diminishing marginal productivity of capital at an economy-wide level for a given supply of labour. Since in such a case increasing knowledge cannot fully compensate for the diminishing returns to the firm's stock of capital, long-run growth will cease unless the exogenous rate of growth of the labour force (population growth) is sufficiently high. Therefore long-run economic growth still depends on an exogenous factor. Nevertheless, Arrow's model can be referred to as an endogenous growth model *avant la lettre*. It is also to be noted that the AK approach in endogenous growth theory is essentially a straightforward extension of the Harrod–Domar model, 'but with sophisticated bells and whistles' (Solow 1994: 49).
5. Hence the crucial question here is: how long is the long run? In the words of Solow: 'It is one thing to say that a quantity will eventually exceed any bound. It is quite another to say that it will exceed any stated bound before Christmas' (Solow 1994: 50).
6. In addition it has been argued that the relevance of an evolutionary perspective for studying energy–economy–climate interactions moves beyond the issue of technological change, because sustainable development implies irreversible, path-dependent change and long-run mutual selection of environmental and economic processes and systems, suggesting at least a conceptual affinity between evolutionary and environmental economics (Mulder and van den Bergh 2001, van den Bergh and Gowdy 2000).
7. We refer to de Vries (2003) for a recent application of evolutionary (game) theory to issues of technology diffusion and environmental policy.
8. It was Spencer, not Darwin, who coined the phrase 'survival of the fittest' (see Hodgson 1993).
9. This is known as Fisher's Fundamental Theorem of Natural Selection.
10. It is to be noted that, despite its unconventional ('evolutionary') form, the model of Conlisk (1989) can accommodate standard patterns of the neoclassical literature on economic growth. Conlisk shows that because technological change is modelled through shifts in the mean of the productivity distribution, respecification of the productivity measure (that is the mean of the productivity distribution) enables his model to reproduce exogenous technical change as in Harrod–Domar and Solow–Swan (1956), learning-by-doing as in Arrow (1962) and incremental learning-by-doing as in Romer (1986).
11. The system of Lotka–Volterra equations describes predator–prey interactions based on growth equations for the interacting population levels. They can be interpreted as an alternative description of basically the same dynamics described by replicator equations (see also Nijkamp and Reggiani 1998).
12. Edenhofer and Jaeger (1998) have recently applied this approach in the field of energy technologies.
13. It is to be noted that this theory of irreversible investment in an uncertain (stochastic) environment has not yet been included in growth theory.

PART II

Theoretical Models

3. Explaining slow diffusion of energy-saving technologies: returns-to-diversity and learning-by-using in a vintage model[1]

3.1 INTRODUCTION

Many existing energy-efficient and cost-effective technologies are not actually in use. The aim of this chapter is to better understand this phenomenon, known as the energy efficiency paradox (Jaffe and Stavins 1994, Jaffe et al. 1999). The question why firms do not immediately and exclusively invest in the newest technologies has already received much attention in the literature. We can distinguish four major explanations for the relatively slow diffusion of new technologies. The first explanation is that the combination of uncertainty and some degree of irreversibility in investment creates an option-value of waiting (see for example Balcer and Lippman 1984, Dixit and Pindyck 1994, Farzin et al. 1998). The second explanation stresses strategic issues: in a world characterized by spillovers and limited appropriability, the presence of (expected) rival innovation and imitation creates an incentive for firms to postpone innovation or adoption (see for example Kamien and Schwartz 1972, Reinganum 1981). The third explanation highlights the fact that over time the performance of existing technologies improves and their price reduces due to learning-by-doing and spillover effects (see for example Grübler et al. 1999, Jovanovic and Lach 1989, OECD/IEA 2000). A final explanation emphasizes the role of vested interests. As switching to new technologies temporarily reduces expertise and hence destroys rents associated with working with relatively old technologies for particular subgroups in the economy, these groups may engage in efforts aimed at keeping the old technologies in place (see for example Canton et al. 2002, Krusell and Ríos-Rull 1996, Mokyr 1992).

In this chapter we offer two additional explanations for the slow diffusion of energy-saving technologies. The first is rooted in a complementarity effect and the second in a learning-by-using effect associated with the use of the technology. We argue that complementarity (or, alternatively, imperfect substitutability) is not only a by-product of past investment decisions, but also an essential ingredient of the process of technological change. It is evident that at the macro level there is continuous investment in both old and new technologies. Similar patterns exist at the sector or even at the firm level, depending on the technology and the type of production process. Many new technologies pass through a life cycle in which they initially complement older technologies, and only subsequently (and often slowly) substitute for older technologies. A number of historical examples, for example the replacement of the waterwheel by the steam engine or the diffusion of new types of processes in the iron and steel industry, illustrate the relevance of complementarities in this 'life cycle view' of technological change (see for example Rosenberg 1976, 1982; Young 1993b). One can argue that modern production processes consist of even more interrelated and mutually reinforcing technologies than the documented historical examples. Hence the production process may be seen as a puzzle of a large number of interdependent and thus complementary technology pieces, making it profitable for firms to continually invest in improvement of distinct pieces instead of replacing the whole puzzle at once (see for example Antonelli 1993, Antonelli et al. 1990, Berg and Friedman 1977, Colombo and Mosconi 1995, Jovanovic and Stolyarov 2000, Milgrom et al. 1991). In sum, the complementarity effect emphasizes the multidimensional character of quality. Technologies differ not only in terms of their productivity (the vertical dimension), but also with respect to other qualities as a result of which firms face returns to diversity (the horizontal dimension). A good example of the latter can be found in the electricity sector where hydropower, unlike thermal power, is characterized by relatively high fixed costs (system of water-inflow and reservoir capacity) and low variable costs (expanding output capacity). Von der Fehr and Sandsbråten (1997) argue that because of this quality (and given the fact that output can only be stored at high costs or not at all), efficiency requires the use of a mix of hydro and thermal technologies, of which hydro is used in particular to meet the need for peak-load capacity. Other benefits from using complementary technologies arise for example from flexibility with respect to inputs (different technologies use different types of fuels or raw materials – for example, iron ore and scrap in steel production), plant location, plant size, or required managerial and organizational skills (Rosenberg 1982). The model that we develop in this chapter captures these phenomena in a stylized way.

The slow diffusion of technologies that is associated with the preference for technological diversity can be further intensified by learning-by-using effects, our second explanation for the slow diffusion of energy-saving technologies. In accordance with broad historical evidence (for example Mokyr 1990, Rosenberg 1982, Young 1993a) we allow for new technologies to be initially inferior to more mature technologies (along the vertical dimension). Learning-by-using improves the productivity of the new technology over time. Therefore, a switch of technologies temporarily reduces expertise and thereby the productivity level of the capital stock. The prospect of such a temporary productivity drop prevents agents from immediate and 'total' switching, but instead induces a gradual adoption of new technologies resulting in coexistence of old and new technologies.

We explore the above-mentioned ideas by developing a simple two-sector macroeconomic model including a final goods sector, producing a homogenous consumption good, and a capital production sector producing heterogeneous vintages. The distinctive features of our model are as follows.[2] First, technology is embodied in physical capital. Newer vintages need less energy and labour to produce the same output than older vintages. This reflects the vertical dimension of quality. Second, different vintages are imperfect substitutes in production. The economy exhibits a 'taste for diversity' of vintages, creating an incentive to invest in both new and older technologies. This reflects the horizontal dimension of quality. Third, our model allows for the endogenous determination of the number of vintages used in the final goods sector, so we offer an economically motivated approach for the scrapping of vintages. Fourth, the representative firm in the final goods sector gains expertise in a technology by using the technologies in its production process. In other words, we include learning-by-using.[3]

There are a number of related articles in which issues of learning and technological innovation and diffusion are analysed. Without extensive discussion we refer to, for example, Aghion and Howitt (1996), Aghion et al. (1997, 1999), Arrow (1962), Chari and Hopenhayn (1991), Jovanovic and Nyarko (1996), Parente (1994), Stokey (1988) and Young (1993a, b). We refer to Jaffe et al. (2002) for an extensive survey of the literature on environmental or energy-saving technological change. The main differences between these articles and our volume lie in our focus on energy-saving technological change, the specifications we make with respect to the complementarity of vintages, the interpretation of intermediate goods as technologies, the emphasis on the complementarity effect in diffusion processes instead of innovation processes (cf. Young 1993b) and the endogenous scrapping mechanism. In focusing on diffusion and adoption, the analysis in this chapter differs from most literature on induced or endogenous technological change, for which the focus is almost solely on innovation (see for example Goulder and Schneider 1999, Goulder and Mathai 2000).

The chapter is organized as follows. In section 3.2 we set up the basic model. In section 3.3 we derive the solution of the model. Section 3.4

illustrates the properties of the model by providing some comparative statics on the complementarity effect and the learning-by-using effect. Section 3.5 brings together the gained insights to address energy-saving technological change in more detail, including an analysis of the effects of imposing an energy price tax. Section 3.6 concludes.

3.2 THE MODEL

The model that we develop is a two-sector vintage model that is characterized by returns to diversity and learning-by-using. The two sectors that we distinguish are (1) a final goods sector in which a homogeneous consumption good is produced using labour, capital and energy, and (2) a capital goods sector consisting of T monopolistically competitive firms each producing a unique vintage of capital using labour. Energy is a complement to capital. We assume that there is one representative consumer, maximizing a simple standard linear utility function subject to a standard budget constraint, for given prices and initial endowments. Utility is a function of consumption of the final good. The income of the consumer is given by the value of his endowments, that is, by the value of his inelastically supplied labour and by the profits of the capital goods sector (which are zero in equilibrium). The inelastic labour supply is equal to L.

The model that we develop can be considered as a vintage model in the sense that capital inputs used in production are heterogeneous in their productivity or quality. For simplicity, capital is assumed to be non-durable. The model can therefore also be considered as a model with heterogeneous intermediate inputs.[4] Technological change results in increased overall productivity and increased energy productivity. Since technologies are embodied in physical capital, newer vintages (or intermediates) are more productive and more energy-efficient than older vintages. We have called this the vertical dimension of quality. The vintages are imperfect substitutes, as discussed in the introduction, and hence there are returns to diversity: for the production of final goods a number of vintages is used. We have called this the horizontal dimension of quality. The formulation that we use in our model to capture this idea is inspired by the product-variety theory which started with the seminal work of Dixit and Stiglitz (1977) and was later extended and applied by, for example, Ethier (1982), Grossman and Helpman (1991) and Romer (1990).

Figure 3.1 illustrates these two dimensions of quality in a stylized way. On the vertical axis is the productivity of the individual vintages. On the horizontal axis are the different vintages that are simultaneously in use because of the returns to diversity. They are ordered according to the date at which they became available.

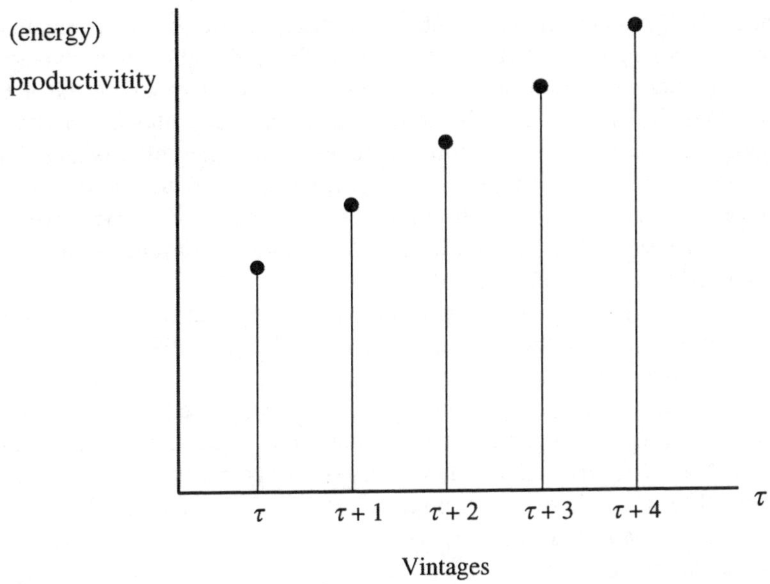

Figure 3.1 Vertical versus horizontal dimension of vintage quality

The Final Goods Production Sector

The final goods sector produces a homogeneous consumption good according to a Cobb-Douglas production function:

$$Y_t = K_t^{\alpha} L_{Yt}^{1-\alpha} , \tag{3.1}$$

in which Y_t represents output produced at date t, and K_t and L_{Yt} are the capital and labour input in final goods production, respectively.[5] Energy is assumed to be complementary to capital (we return to this below). Capital is an aggregate of vintages of capital goods. Vintages are characterized by the first date of their availability τ.

Following the seminal work of Dixit and Stiglitz (1977), we specify the aggregate capital stock as:

$$K_t = \left[\int_{t-T}^{t} \left(A_{\tau,t} K_{\tau,t} \right)^{\frac{\varepsilon-1}{\varepsilon}} d\tau \right]^{\frac{\varepsilon}{\varepsilon-1}} \quad (\varepsilon > 1) \;, \tag{3.2}$$

in which T is the (endogenous) number of vintages in use, $K_{\tau,t}$ is the amount of capital of vintage τ used at date t (where $t-T \leq \tau \leq t$) and $A_{\tau,t}$ is a vintage-specific productivity parameter. Alternatively T can be interpreted as the age of the oldest vintage in use. Technological change is embodied in new vintages. The elasticity of substitution between any pair of vintages (in efficiency units) is denoted by ε. Vintages are assumed to be imperfect substitutes ($\varepsilon > 1$). This formulation implies that firms have an incentive to invest in older technologies, even if new technologies are available that are characterized by a higher productivity A.

It is a common characteristic in vintage modelling that it is optimal for firms to invest only in the newest technology (Meijers 1994).[6] The very reason that old and new vintages coexist in most vintage models is that once firms have incurred the (partly) sunk investment cost, it need not be optimal to replace existing capital goods once a new technology arrives. According to equation (3.2), however, old and new vintages coexist because of a taste for diversity. The strength of this complementarity effect is defined by the substitution elasticity parameter ε. The larger ε, the lower the degree of complementarity.

The productivity of vintages (that is, the vertical dimension of quality) is affected by two forces. The first is exogenous. As we do not elaborate on the innovation process in the 'intermediate capital' sector, but rather focus on the diffusion of the new vintages in the final good sector, we just assume that newer vintages – as they are brought onto the market – are more productive than older vintages when those were brought onto the market. The second factor is endogenous: vintages improve as they are used. Hence productivity endogenously depends on the cumulative past investments in vintages. We further label this learning-by-using. More specifically, we assume that the productivity of vintages ($A_{\tau,t}$) develops according to

$$A_{\tau,t} = A_0 e^{g\tau} + \left[1 - (1 + aC_{\tau,t})^{\lambda-1} \right] \left(A_\tau^{\max} - A_0 e^{g\tau} \right). \tag{3.3}$$

In this specification, A_0 is initial productivity, g is an exogenously given growth rate of the productivity of new vintages, a measures the strength of learning-by-using effects, C_t represents past cumulative investments in vintage τ, λ represents the curvature of the learning curve, and A_τ^{max} is the vintage-specific maximum productivity level (that is, the prevailing productivity level when the technology has matured). For simplicity, we assume that A_τ^{max} is in fixed proportion $\gamma(\geq 1)$ to the productivity at the date of introduction of the vintage ($A_\tau^{max} = \gamma A_0 e^{g\tau}$). In the special case in which $\gamma = 1$, the learning-by-using mechanism is absent and productivity of vintages purely depends on exogenous improvements. The assumption that $0 < \lambda < 1$ implies that the productivity of a technology in the presence of learning-by-using ($\gamma > 1$) gradually converges to the mature productivity level A_τ^{max}. For the time being, we assume for reasons of analytical tractability that learning-by-using is absent (γ=1). We generalize the specification for the productivity development and discuss the implications of learning-by-using for diffusion patterns and adoption of technologies in section 4.2.

As already mentioned, the energy requirement of a particular vintage at a point in time is directly related to the use of capital. The empirical literature has convincingly shown that substitution possibilities between energy and capital are limited (see for example Berndt and Wood 1979, Hudson and Jorgenson 1974, Kuper and Van Soest 2003, Prywes 1986).[7] For reasons of analytical tractability, we make the extreme assumption of no substitution possibilities:

$$E_{\tau,t} = \psi_\tau K_{\tau,t}, \quad (\psi_\tau > 0), \tag{3.4}$$

where $E_{\tau,t}$ is the energy input for vintage τ at time t and ψ_τ is the vintage-specific energy–capital ratio.[8] The vintage-specific energy–capital ratio is assumed to decline at an exogenously given rate f:

$$\psi_\tau = \psi_0 e^{-f\tau}, \quad (f \geq 0), \tag{3.5}$$

Thus newer vintages are more energy-efficient than older vintages due to exogenous technological progress. Total energy use in the economy equals.

Producers in the final goods sector operate under perfect competition and a representative firm maximizes profit π:

$$\pi_t = P_{Yt} Y_t - w_t L_{Yt} - \int_{t-T}^{t} p_{K\tau,t} K_{\tau,t} d\tau - \int_{t-T}^{t} p_{Et} E_{\tau,t} d\tau, \tag{3.6}$$

in which P_Y, w, $P_{K\tau}$ and P_E denote the output price, the wage rate, the price of capital goods of a specific vintage and the energy price, respectively (we omit time-indices if possible). In the remainder we assume a small open economy, implying that the domestic use of foreign goods is negligible compared with the foreign use of foreign goods. There is only one foreign good in our model that is imported, namely energy. Hence the price of energy is fixed (at the level of the world market prices). Finally, we assume that the final good can be exported and that balanced trade holds. The wage rate is taken to be the numeraire of the model ($w = 1$). The vintage capital needed to produce final goods is bought from the capital goods sector, to which we turn in the next subsection.

The Capital Production Sector

The capital production sector consists of a number of T monopolistically competitive firms each producing a specific vintage according to

$$K_{\tau,t} = L_{K\tau,t}. \tag{3.7}$$

We assume that in each period, a new vintage becomes available due to an exogenous process of technological innovation and only one firm acquires the right to produce capital of this particular vintage.[9] In addition, in each period firms in this sector have to pay a fixed cost in terms of labour (L_f) before being able to produce. Monopoly rents have to compensate for these costs. Firms maximize their profits according to

$$\max \pi_{\tau,t} = p_{K\tau,t} K_{\tau,t} - (L_{K\tau,t} + L_f)w. \tag{3.8}$$

These rents constitute income for the representative consumer. We assume a constant and exogenous labour supply L. The model is closed by imposing labour market equilibrium

$$L = L_{Yt} + \int_{t-T}^{t} (L_{K\tau,t} + L_f) d\tau, \tag{3.9}$$

and by the standard budget constraint of the representative consumer.[10] In the next section we discuss the solution of the model, focusing on the allocation of labour and the determination of the number of vintages used in the production process.

3.3 SOLUTION OF THE MODEL

Final Goods Sector

Producers in the final goods sector solve a standard profit-maximization problem in two stages. First, they determine the optimal relative demand for the composite of capital, energy, and labour by maximizing the profit function (equation 3.6). Using the expression for the use of energy (equation 3.4), this results in the standard allocation rule for a Cobb-Douglas production function implying constant cost shares of capital, including energy costs, and labour:

$$\frac{P_K K_t + p_E \psi K_t}{wL_Y} = \frac{\int_{t-T}^{t} (p_{K\tau} + p_E \psi_\tau) K_\tau d\tau}{wL_Y} = \frac{\alpha}{1-\alpha}, \quad (3.10)$$

in which P_K is the price index of the composite capital good.

Second, producers decide on the optimal amount of each vintage by solving the following maximization problem:

$$\max_{K_{\tau,t}} \left[\int_{t-T}^{t} (A_\tau K_\tau)^{\frac{\varepsilon-1}{\varepsilon}} d\tau \right]^{\frac{\varepsilon}{\varepsilon-1}}$$
$$(3.11)$$
$$s.t. \int_{t-T}^{t} (p_{K\tau} + p_E \psi_\tau) K_\tau d\tau \leq P_{Kt} K_t + p_{Et} E_t \ .$$

Optimization yields a downward-sloping demand curve for capital of a specific vintage:

$$K_\tau = K_s \left[\frac{A_\tau}{A_s} \right]^{\varepsilon-1} \left[\frac{(p_{K\tau} + p_E \psi_\tau)}{(p_{Ks} + p_E \psi_s)} \right]^{-\varepsilon} . \quad (3.12)$$

The relative demand for two vintages of different age (s and τ) thus depends on their relative productivity and their relative prices including energy costs. The 'real' or effective energy costs are determined by the (exogenous) energy price and the energy–capital ratio. The relative demand for a newer vintage τ increases if the substitution elasticity parameter ε increases (recall that $\varepsilon > 1$). The intuition behind the latter is that a lower degree of complementarity implies firms to be more responsive to productivity differences.

Capital Production Sector

Producers in the vintage production sector maximize their profits (equation 3.8) subject to a downward-sloping demand curve for the vintage that they produce (equation 3.12). This results in standard mark-up pricing, according to which the producers of vintages put a mark-up over labour costs and a factor including vintage-specific energy costs:

$$\frac{\partial \pi_\tau}{\partial L_{K\tau}} = 0 \Leftrightarrow p_{K\tau} = \frac{\varepsilon}{\varepsilon - 1} w + \frac{p_E \psi_\tau}{\varepsilon - 1}. \qquad (3.13)$$

This mark-up is larger the greater the complementarity between different vintages (that is, the smaller ε) and the larger the vintage specific energy costs of the firms in the final goods sector (that is, the larger $P_E\psi_\tau$). This basically concludes the description of the behaviour of firms in our economy.

Number of Vintages and Labour Allocation

The model is solved by determining the number of vintages that can be sustained in the economy (that is, the age of the oldest vintage that can be sustained). To understand this intuitively, recall that on the one hand newer vintages are characterized by a higher relative productivity and lower energy costs as a result of which the relative demand for old vintages will gradually decline over time. On the other hand, the complementarity between vintages of different age prevents firms from immediately shifting to the most productive vintage. At some point in time however the demand for a vintage becomes so low that the vintage can no longer profitably be supplied by the producer of that vintage (given the fixed cost in terms of labour that he has to incur). Supply will stop and the vintage disappears from the market. This 'scrapping' of vintages is – in contrast with the more traditional vintage literature – due to it no longer being possible to profitably supply the vintage, whereas the traditional vintage literature explains scrapping from the fact that at some point in time, the vintage can no longer profitably be used by the owner (see for example den Hartog and Tjan 1980, Malcolmson 1975).

To determine the number of vintages used, we first need to determine the allocation of labour over the production or assemblage of final goods and the production of vintages, respectively. Using the fact that cost-shares of the inputs are constant, we can determine the allocation of labour. For the time being, we assume for reasons of analytical tractability that the energy–capital ratio ψ_τ is constant for all vintages ($\psi_\tau = \psi$). We generalize and discuss the implications of allowing for a vintage-specific energy–capital ratio for diffusion patterns and adoption of technologies in section 3.5.

Using equations 3.7, 3.10 and 3.13, we derive the allocation of labour as (see Appendix 3.A):

$$L_Y = \frac{(1-\alpha)}{\alpha}\frac{\varepsilon}{\varepsilon-1}\left(1+\frac{p_E\psi}{w}\right)\int_{t-T}^{t} L_{K\tau}d\tau . \qquad (3.14)$$

This expression reveals that more assemblage labour will be used relative to labour used for producing vintages, the smaller the share parameter in the production function of final goods, the higher the energy price and the energy–capital ratio, and the lower the elasticity of substitution. Both a lower elasticity of substitution and an increase in the costs of energy result in relatively high prices of vintages due to mark-up pricing (equation 3.13). Therefore a shift from capital to labour in final goods production will take place. Substituting this equation into the definition for the labour market equilibrium (equation 3.9) and rewriting yields the expression for the labour used to produce the vintage capital stock (see Appendix 3.A):

$$\int_{t-T}^{t} L_{K\tau}d\tau = \frac{\alpha(\varepsilon-1)w}{(\varepsilon-\alpha)w+(1-\alpha)\varepsilon p_E\psi}\left(L-\int_{t-T}^{t} L_f d\tau\right). \qquad (3.15)$$

Firms in the capital production sector continue producing their specific vintage as long as they are compensated for their production costs. So they produce as long as

$$p_{K\tau}K_\tau \geq (L_{K\tau}+L_f)w . \qquad (3.16)$$

Using the production function for vintages and mark-up pricing (equations 3.7 and 3.13), this expression can be rewritten (with equality) as

$$\frac{\varepsilon}{\varepsilon-1}=\frac{L_{K\tau}+L_f}{L_{K\tau}}-\frac{p_E\psi}{(\varepsilon-1)w}. \tag{3.17}$$

This expression basically determines the minimum required scale of operation for a producer of vintages, and hence the minimum demand for a particular vintage that is needed for the producer of that vintage to be able to operate profitably. From equation (3.17) and using equation (3.7) this minimum demand can be derived as

$$\overline{K}=\overline{L}=\left(\frac{\varepsilon-1}{\frac{p_E\psi}{w}+1}\right)L_f, \tag{3.18}$$

in which $\overline{K} = K_{t-T,t}$ is the amount of the oldest vintage that will be produced and $\overline{L} = L_{t-T,t}$ is labour used for its production. The minimum scale of operation or the minimum demand for a particular vintage is larger the higher the fixed cost, the lower the energy–capital ratio and the higher the elasticity of substitution (and hence the lower the mark-up the producers of the vintages can charge). Any firm that produced an older vintage, for which there would be less demand due to its lower productivity, would make losses.

Having determined the production level of the oldest vintage, we can uniquely determine the production levels of more recent vintages that are in use by combining the expression for the relative demand for different vintages and the productivity difference between these vintages. Substituting the expressions for the capital price (equation 3.13) into equation (3.12) and rewriting yields (see Appendix 3.B):

$$L_{K\tau,t}=\overline{L}e^{g(\varepsilon-1)(\tau+T-t)}. \tag{3.19}$$

Figure 3.2 graphically illustrates this expression by displaying the production of one particular vintage (arriving on the market at t = τ) over time. This expression reveals that in the presence of exogenous improvements of the performance of newer vintages ($g > 0$), more labour is used for the production of more recent vintages. This effect is reinforced when the degree of complementarity among vintages declines (ε increases).

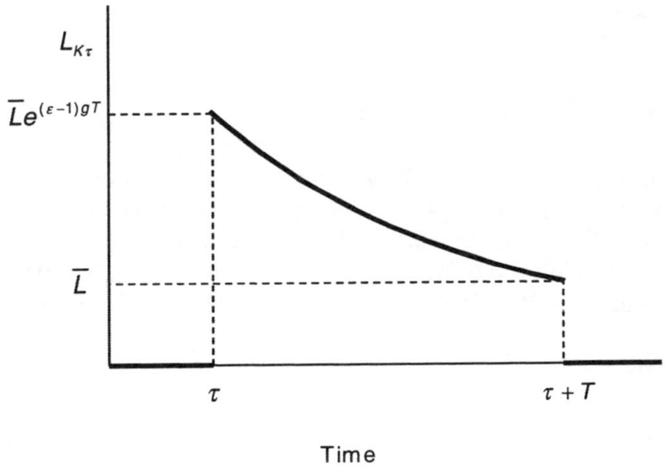

Figure 3.2 Demand for one vintage during its lifetime

The total amount of labour used for the production of vintages equals (using equations 3.18 and 3.19):

$$\int_{t-T}^{t} L_{K\tau,t}d\tau = \int_{0}^{T} \overline{L}e^{(\varepsilon-1)g\tau}d\tau = \frac{\overline{L}\left(e^{(\varepsilon-1)gT}-1\right)}{g(\varepsilon-1)} = \frac{L_f\left(e^{(\varepsilon-1)gT}-1\right)}{g\left(\frac{p_E\psi}{w}+1\right)}. \quad (3.20)$$

Combining equations (3.15) and (3.20) solves the model for the mass of vintages that can be sustained in the economy (or alternatively the age of the oldest vintages in use). This solution for T is given by the following implicit function:

$$((\varepsilon-\alpha)w+(1-\alpha)\varepsilon p_E\psi)L_f(e^{(\varepsilon-1)gT}-1)- \\ \alpha g(\varepsilon-1)(p_E\psi+w)(L-TL_f)=0. \quad (3.21)$$

In the next section we illustrate the solution and the comparative statics of the model, focusing on the two explanations we offered in our model for the slow diffusion of new technologies: the complementarity effect and the learning-by-using effect. In section 3.5 we apply these insights to study energy-saving technological change in more detail and to analyse the effects of imposing an energy price tax in the presence of technological complementarities.

3.4 COMPARATIVE STATICS

The aim of this section is to illustrate the properties of our model by providing some comparative-static characteristics of the model. First, we analyse the role that complementarities between different technologies play in explaining adoption and diffusion of new vintages. Second, we elaborate on the role of learning-by-using.

Complementarity

The degree of complementarity is captured by the elasticity of substitution between the vintages. The consequences of a higher degree of complementarity (that is, a lower elasticity of substitution) can best be understood by dividing the total effect in three components, which we summarize in proposition 1.

Proposition 1. An increase in the degree of complementarity between vintages implies (1) a decrease of the minimum demand required for vintage producers to operate profitably, (2) a lower responsiveness of the relative demand for vintages to increases in productivity of newer vintages, and (3) a decrease in the demand of firms in the final goods sector for capital. As a result diffusion of new vintages slows down.

These three effects are illustrated in Figure 3.3.[11] The figure depicts the demand for vintages of different age (on the vertical axis) as a function of the date of introduction on the market (on the horizontal axis). The most recent (current) vintage is located most to the right in the figure. The figure can be understood as follows. Consider first the case in which the elasticity of substitution is high (the solid line). The upward slope of the line reflects the fact that newer vintages (located more to the right) have a higher productivity and consequently have a higher demand. The demand of the oldest vintage is given by the minimal required demand as defined in equation (3.17) and is represented by the lowest point on the bold line. The area under this line is equal to the amount of labour that is available for the production of vintages as it is given in equation (3.14). Combining these three elements yields a unique solution of the model that is essentially characterized by the age of the oldest vintage.

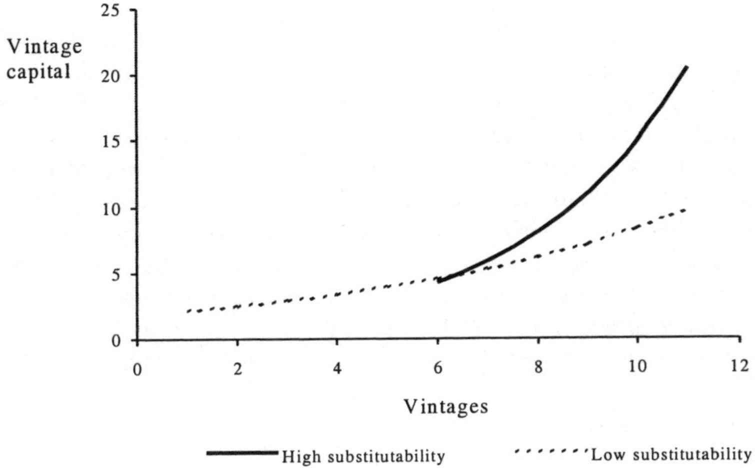

Figure 3.3 Demand for different vintages in absence of learning

Let us now consider what happens when the degree of complementarity increases (that is, the elasticity of substitution decreases). First, for each producer of vintages the minimum required demand to operate profitably decreases because the producer can charge a higher mark-up (see equation 3.13). This implies *ceteris paribus* an increase in the equilibrium number of vintages sustained in the economy. Second, the demand curve becomes flatter as the relative demand for vintages becomes less responsive to productivity differences (see equation 3.12). *Ceteris paribus* this also implies that the equilibrium number of vintages that can be sustained in the economy increases. Third, producers of final goods shift towards labour as this becomes cheaper relative to capital. This is reflected by a downward shift of the demand curve. This effect opposes the previous two effects and implies that less vintages can be sustained. It can be shown however that the former two effects dominate for reasonable parameter values,[12] so a higher degree of complementarity increases the number of vintages that can be sustained. In other words the existence of complementarity slows down the rate of modernization of the capital stock.

Learning-by-Using

In the previous section we have assumed a learning effect to be absent for reasons of analytical tractability. In this subsection we include learning-by-using into the model according to the specification given in equation (3.3). This implies that the productivity of a vintage increases over time as a function of past cumulative investment in that vintage. The productivity improvement is initially relatively large. Furthermore it is bounded in the sense that there is a vintage-specific maximum productivity level. This formulation captures the available empirical evidence that productivity increases at a relatively fast rate just after a technology is introduced, in order to slow down at later stages, and to level off when the technology matures (Grübler et al. 1999). From this specification, we can derive proposition 2.

Proposition 2. Learning-by-using implies that the most productive vintage is of an intermediate age and hence initial demand for the newest vintage is relatively low. As a result, diffusion of new vintages slows down.

A relatively strong initial learning rate implies that situations can arise in which the productivity of vintages introduced some periods ago exceeds the productivity of vintages that have just been introduced. Figure 3.4 depicts a typical example of this kind of productivity development for two different vintages.

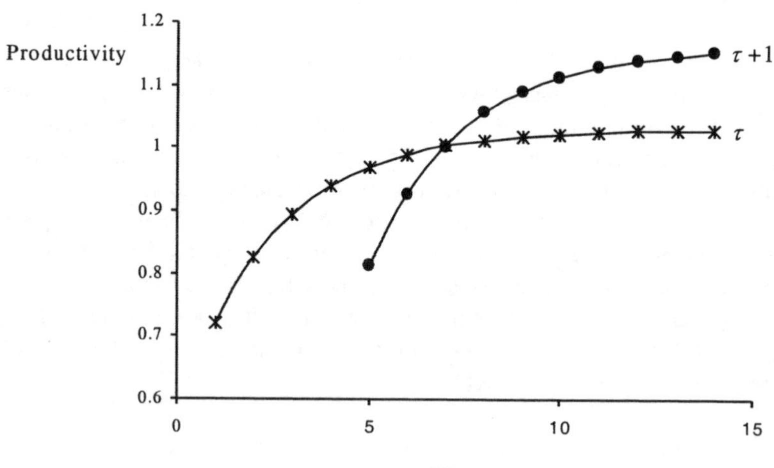

Figure 3.4 Productivity development of different vintages with learning-by-using

The newer vintage (starting more to the right) is potentially more productive as it comes on the market (at $t = \tau$), but initially the old technology outperforms the new technology due to a learning-by-using effect. The productivity development reflected in Figure 3.4 implies the possibility (dependent on the optimal number of vintages) of the most productive vintage being a vintage of intermediate age. Older vintages are less productive since they have matured, whereas newer vintages have not yet fully reaped the productivity improvements that arise from learning-by-using.

The implications of such developments for the diffusion of new technologies are illustrated in Figure 3.5, which is similar to Figure 3.3, but shows the case with learning-by-using.[13]

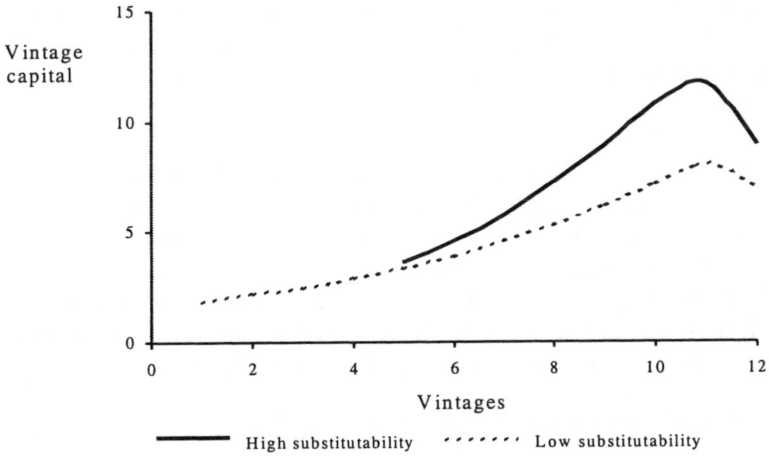

Figure 3.5 Demand for different vintages with learning-by-using

New vintages are initially demanded at a relatively limited scale due to their low productivity, but as they improve due to learning-by-using, demand will increase, and subsequently they will be gradually phased out of the production process as the vintage matures and ultimately becomes obsolete.

The positive effect of learning-by-using on the productivity of older vintages provides a barrier to adopting newer vintages and hence slows down the diffusion process. Based upon the logic as explained earlier in this section, a lower elasticity of substitution results in more vintages being used in the production process, and lower responsiveness to differences in productivity levels between vintages of different age. In other words, in the presence of complementarity between vintages, learning-by-using intensifies the slow rate of modernization of the capital stock.

3.5 ENERGY-SAVING TECHNOLOGICAL CHANGE AND ENERGY TAX POLICIES

In this section we bring together the insights on the role of technological complementarities to address the issue of energy-saving technological change in more detail. First we provide a numerical analysis of energy-saving technological change. Then we analyse the impact of imposing an energy price tax in the presence of technological complementarities.

Dynamics of Energy-Saving Technological Change

First, we compare two firms or sectors that differ from each other in terms of energy efficiency. All other things being equal, we analyse our model for two constant values of the energy–capital ratio parameter ψ. In other words the two firms or sectors differ in terms of energy efficiency, but for each firm the energy efficiency is constant over all vintages. We summarize the results in proposition 3.

Proposition 3. A relatively high (low) energy intensity implies (1) a lower (higher) responsiveness of the relative demand for vintages to a higher price of capital, (2) a lower (higher) minimum demand required for vintage producers to operate profitably, and (3) a lower (higher) demand of firms in the final goods sector for capital. As a result diffusion of new vintages is relatively slow (fast). Exogenous energy-saving technological improvement of newer vintages speeds up modernization of the capital stock.

An illustration is given in Figure 3.6, showing the demand for vintages for a comparative static situation with a relatively high and a relatively low level of energy intensity. Since energy is complementary to capital (equation 3.4) a higher ψ implies that the sector or firm requires a relatively high amount of energy to produce final goods, which in turn means that energy costs comprise a significant part of the costs of the capital–energy composite in the final goods sector. Vice versa this means that the price of vintage capital is of relatively less weight in determining investment decisions of firms, as a result of which the demand for vintages reacts only moderately to an increase in capital price. This induces monopolistic producers of vintage capital to set a relatively high mark-up (equation 3.13) which implies that they need less demand for their vintage to compensate for their fixed costs (equation 3.18). Hence the economy or firm will sustain a relatively large number of vintages whereas the demand for each vintage is relatively low since the capital–energy composite is relatively expensive due to the high energy intensity.[14] This is illustrated in Figure 3.6 with the dashed line.

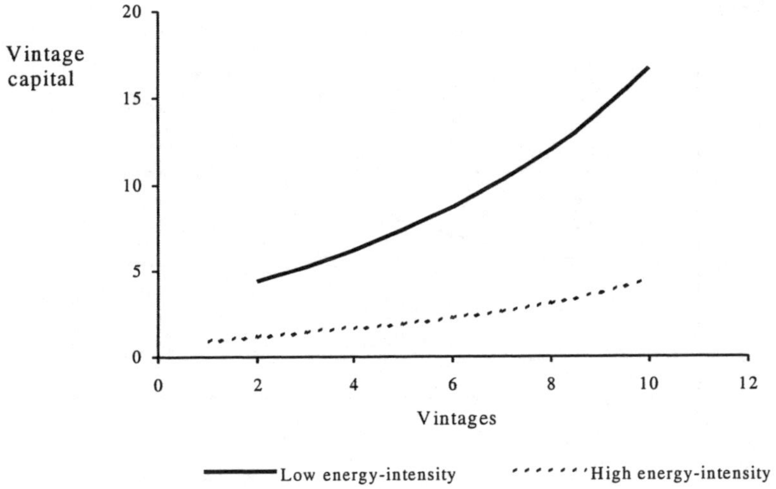

Figure 3.6 Demand for different vintages with two levels of energy-intensity (in the absence of learning)

The bold line in Figure 3.6 illustrates the low energy intensity case. The argument is now the other way around: if a firm is characterized by relatively low energy intensity, the price of capital forms a relatively large part of the cost of the capital–energy composite. This implies that the demand for vintages is relatively sensitive to a change in capital price, reducing the market power of the producers of vintage capital and subsequently the mark-up they can set. Hence the firm will sustain a relatively small number of vintages, whereas the demand for each vintage is relatively high since the capital–energy composite is relatively cheap due to the low energy intensity. As a consequence a relatively large amount of labour is used to produce vintage capital (see equations 3.14 and 3.15). This is reflected by an upward shift of the demand curve for vintages.

Next we analyse what happens to a firm when a new energy-saving technology arrives on the market.[15] Of course the firm will invest in the new technology since the new technology is better, *ceteris paribus*. Because of the complementarity effect however, firms continue to invest in old technologies as well and we thus have a transition period. Figure 3.7 shows a firm's demand for vintages at a particular point in time after the (exogenous) arrival of a new energy-saving technology at time $t = S$.

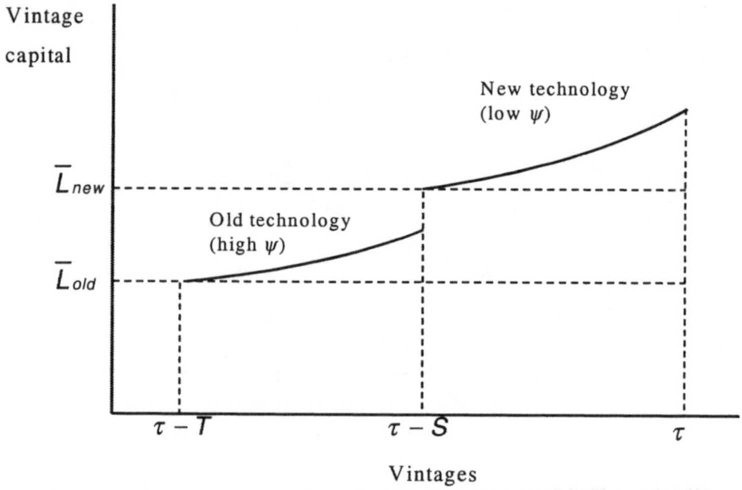

Figure 3.7 Demand for vintages during the transition to a new technology

The demand curve for vintages essentially combines the two demand curves depicted in Figure 3.6. The most recent vintages ($\tau > (t - S)$) embody the new energy-saving technology while the older vintages ($(t - T) < \tau < (t - S)$) embody the older technology. As we argued above, an increase in energy efficiency reduces the mark-up set by the producer of a vintage, and hence increases the minimum demand $\overline{L}$ needed to cover the fixed costs. As a result the demand curve for vintage capital is discontinuous during the transition period.

Finally, we relax the assumption of a constant energy–capital ratio (that is we assume $f > 0$) within each sector or firm. This means that for each firm more recent vintages are less energy intensive than older vintages. The amount of labour used for the production of vintages can be derived from the generalized version of equation (3.19) (see Appendix 3.B):

$$L_{K\tau,t} = \overline{L}_t e^{g(\varepsilon-1)(\tau+T-t)} \left[\frac{1+\dfrac{p_E}{w}\psi_0 e^{-f\tau}}{1+\dfrac{p_E}{w}\psi_0 e^{-f(t-T)}} \right]^{-\varepsilon}$$

where $T - t < \tau < t$, $e^{-f\tau} < e^{-f(t-T)}$.

This expression reveals that in the presence of exogenous improvements in the energy efficiency of newer vintages ($f > 0$), more labour is used for the production of more recent vintages (higher τ). This implies that the lines in Figures 3.6 and 3.7 get steeper as compared to the case with a constant energy–capital ratio for all vintages ($f = 0$). Hence, modernization of the capital stock is faster. Moreover, because the capital–energy composite becomes more energy efficient over time, and thus cheaper, the demand for vintage capital increases over time, inducing scope for additional vintages to be sustained in the economy.

Energy Price Tax

Finally, we analyse the consequences of the introduction of an energy tax that given the earlier discussion is easy to understand. We confine ourselves to analysing the effects of an energy price tax assuming constant values of the energy–capital ratio parameter ψ for all vintages within each firm, while they may differ between firms. We summarize the results in proposition 4.

Proposition 4. Imposing an energy price tax under the assumption of a constant energy–capital ratio (ψ) for all vintages within a firm will, ceteris paribus, reduce the amount of energy used and slow down diffusion of new vintages.

The first effect in proposition 4 results from a lower demand for vintage capital due to the increasing costs of the capital–energy composite. Of course this effect depends crucially on the assumption of no substitution possibilities between energy and capital (equation 3.4). The second effect is due to the fact that old vintages will be sustained for a relatively long period since producers of vintages need less demand to compensate for the fixed costs. Again the underlying reason is that they are able to set a higher mark-up because higher energy prices lower the relative weight of capital costs in determining investment decisions of firms, as a result of which the demand for vintages reacts only moderately to an increase in capital price (see equation 3.13).

Of course it is more realistic to allow for (exogenous) improvement of energy efficiency over time, implying an increased demand for more recent vintages, as they are more energy efficient. Imposing an energy price tax then further increases the demand for recent energy-efficient vintages, reducing energy consumption. The total effect on the amount of energy used however is ambiguous, because increasing energy efficiency makes the capital–energy composite cheaper, inducing an increase in the demand for vintage capital and thus energy (see equation 3.4). Which effect dominates depends on the relative magnitude of the energy price tax and the exogenous energy efficiency improvement (*f*).

3.6 CONCLUSIONS

The widespread adoption of new energy-efficient technologies is a lengthy and costly process. In this chapter we developed a vintage model to study the diffusion of energy-saving technologies and to explain why diffusion is gradual and why firms continue to invest in old technologies. An important characteristic of our model is that vintages are complementary – in other words there are returns to diversity of using different vintages. We have argued that this is a potentially relevant part of the explanation of the energy efficiency paradox. Furthermore, we have shown that this effect is intensified when we take learning-by-using effects into account. Since a firm faces loss of expertise on a particular vintage, gained by virtue of experience, when it switches to a newer vintage, the existence of learning-by-using processes can provide an extra argument for firms to continue investing in older vintages. These results illustrate that the impact of learning processes on technological progress is less straightforward than is often assumed in macroeconomic models emphasizing the positive aggregate impact of knowledge creation. Our model has shown that at a microeconomic level, knowledge accumulation might even slow down processes of technological change in the short run, in particular if knowledge spillovers between different vintages (or product varieties) are limited. We refer to Benkard (2000) for an empirical illustration of this point.

In addition our model illustrates that the development of energy consumption depends on the interplay between the rate and the nature of technological change, and the stringency of environmental policy. We have shown that an increase in energy efficiency speeds up scrapping of older technologies and induces an increase in the demand for vintage capital, since it makes the capital–energy composite cheaper. The former reduces the demand for energy whereas the latter increases the demand for energy, since we assume no substitution possibilities between capital and energy. Finally, we have shown that an energy price tax speeds up the diffusion of new energy-saving technologies and induces substitution from capital to labour, both decreasing the demand for energy. Hence the development of aggregate energy consumption depends on the relative magnitude of these different effects.

In this chapter we have argued that slow diffusion of energy-saving technologies is less a paradox than it might seem to be, for it can be the result of rational economic investment decision-making that takes into account the sunk investment costs of installed vintages, returns to diversity and the costs inherent to the loss of expertise. In the next chapter we continue our exploration of the energy efficiency paradox by analysing microeconomic technology adoption decisions under uncertainty, and the role subsidies can play in stimulating energy-saving technological change.

APPENDIX 3.A

In this appendix we determine the allocation of labour over the production of final goods and the production of vintages, respectively. Using the fact that cost-shares of capital and intermediates are constant, we can determine the allocation of labour. This results in an expression for the labour used to produce vintage capital. Recall from equation (3.13) that $p_{K\tau,t} = \frac{\varepsilon}{\varepsilon - 1} w_t + \frac{p_{Et}\psi_\tau}{\varepsilon - 1}$ and from equation (3.7) that $K_{\tau,t} = L_{k\tau,t}$. Substituting equation (3.13) and equation (3.7) into equation (3.10) yields:

$$\frac{\int_{t-T}^{t} \left(\frac{\varepsilon}{\varepsilon - 1} w_t + \frac{p_{Et}\psi_\tau}{\varepsilon - 1} + p_{Et}\psi_{\tau,t} \right) L_{K\tau,t} d\tau}{w_t L_{Yt}} = \frac{\alpha}{1-\alpha}. \tag{3.A.1}$$

Recall that we assumed that the energy price p_{Et} and the mark-up $\varepsilon/(\varepsilon - 1)$ are uniform for all vintages. Furthermore recall that we assume for simplicity that the energy–capital ratio ψ_τ is the same for all vintages ($\psi_\tau = \psi$). So equation (3.A.1) reads as:

$$\frac{\left(\frac{\varepsilon}{\varepsilon - 1} w_t + \frac{p_{Et}\psi_\tau}{\varepsilon - 1} + p_{Et}\psi_{\tau,t} \right) \int_{t-T}^{t} L_{K\tau,t} d\tau}{w_t L_{Yt}} = \frac{\alpha}{1-\alpha}. \tag{3.A.2}$$

Rewriting yields:

$$L_Y = \frac{(1-\alpha)}{\alpha} \frac{\varepsilon}{\varepsilon - 1} \left(1 + \frac{p_E \psi_t}{w} \right) \int_{t-T}^{t} L_{K\tau,t} d\tau, \tag{3.A.3}$$

which corresponds to equation (3.14). Substituting this equation into the definition for the labour market equilibrium (equation 3.9) yields:

$$L = \left[\frac{(1-\alpha)}{\alpha} \frac{\varepsilon}{\varepsilon - 1} \left(1 + \frac{p_E \psi_t}{w} \right) + 1 \right] \int_{t-T}^{t} L_{K\tau,t} d\tau + \int_{t-T}^{t} L_f d\tau . \quad (3.A.4)$$

Rewriting yields the expression for the labour used to produce the capital stock:

$$\int_{t-T}^{t} L_{K\tau,t} d\tau = \frac{\alpha(\varepsilon - 1)w}{(\varepsilon - \alpha)w + (1-\alpha)\varepsilon p_E \psi_t} \left(L - \int_{t-T}^{t} L_f d\tau \right), \quad (3.A.5)$$

which corresponds to equation (3.15) (we omit time indices in the main text).

APPENDIX 3.B

In this appendix we derive an expression that relates the production of each vintage to the production of the oldest vintage $\overline{K} = K_{t-T,t}$. Substituting equation (3.13) into equation (3.12) yields:

$$K_\tau = K_s \left[\frac{A_\tau}{A_s} \right]^{\varepsilon - 1} \left[\frac{w + p_E \psi_\tau}{w + p_E \psi_s} \right]^{-\varepsilon} . \quad (3.B.1)$$

If we assume index *t-T* to denote the oldest vintage in use and substitute it for the vintage with index *s*, equation (3.B.1) develops into:

$$L_{K\tau} = K_{t-T} \left[\frac{A_\tau}{A_{t-T}} \right]^{\varepsilon - 1} \left(\left[\frac{w + p_E \psi_\tau}{w + p_E \psi_{t-T}} \right]^{-\varepsilon} \right). \quad (3.B.2)$$

Recall that $K_{\tau,t} = L_{K\tau,t}$ (equation 3.7) and that $\psi_\tau = \psi_0 e^{-f\tau}$ (equation 3.5). Furthermore recall that we assumed absence of learning, so $A_{\tau,t} = A_0 e^{g\tau}$ (see equation 3.3). Substituting these expressions into equation (3.B.2) yields:

$$L_{K\tau} = L_{t-T} \left[\frac{A_0 e^{g\tau}}{A_0 e^{g(t-T)}} \right]^{\varepsilon - 1} \left[\frac{w + P_E \psi_0 e^{-f\tau}}{w + P_E \psi_0 e^{-f(t-T)}} \right]^{-\varepsilon} . \quad (3.B.3)$$

Rewriting yields:

$$L_{K\tau} = L_{t-T} e^{g(\varepsilon-1)(\tau+T-t)} \left[\frac{w + P_{Et}\psi_0 e^{-f\tau}}{w + P_{Et}\psi_0 e^{-f(t-T)}} \right]^{-\varepsilon} . \tag{3.B.4}$$

From equation (3.18) it can be seen that the labour used to produce the oldest vintage, $L_{Kt-T,t}$, equals the minimum demand needed for the producer of that vintage to be able to operate profitably, that is $\overline{K} = \overline{L}$. As a result the demand (production) for each vintage can be related to the demand (production) for the oldest vintage in the market according to:

$$L_{K\tau,t} = \overline{L}_t e^{g(\varepsilon-1)(\tau+T-t)} \left[\frac{w + P_E\psi_0 e^{-f\tau}}{w + P_E\psi_0 e^{-f(t-T)}} \right]^{-\varepsilon} . \tag{3.B.5}$$

Rewriting yields:

$$L_{K\tau,t} = \overline{L}_t e^{g(\varepsilon-1)(\tau+T-t)} \left[\frac{1 + \frac{P_E}{w}\psi_0 e^{-f\tau}}{1 + \frac{P_E}{w}\psi_0 e^{-f(t-T)}} \right]^{-\varepsilon} , \tag{3.B.6}$$

where $T - t < \tau < t$, $e^{-f\tau} < e^{-f(t-T)}$.

When we assume a constant energy–capital ratio ($\psi_\tau = \psi$), that is, $f = 0$, the second term on the right-hand side of equation (3.B.6) becomes 1 and hence equation (3.B.6) reduces to:

$$L_{K\tau,t} = \overline{L} e^{g(\varepsilon-1)(\tau+T-t)} . \tag{3.B.7}$$

This corresponds to equation (3.19) in the main text.

NOTES

1. This chapter is a slightly revised version of the publication: Mulder, P., H.L.F. de Groot and M.W. Hofkes, 2003, 'Explaining Slow Diffusion of Energy-Saving Technologies: A Vintage Model with Returns to Diversity and Learning-by-Using', *Resource and Energy Economics*, 25, 105–26. Reprinted with permission from Elsevier.
2. The model presented in this chapter essentially extends the basic framework developed in de Groot et al. (2003). The extension consists of, among others, incorporating energy as a

production factor into the model.

3. As noted in Chapter 1, we follow Rosenberg (1982: 121–2) in distinguishing three basic types of learning: learning in R&D stages, learning at the manufacturing stage and learning as a result of use of the product or technology. We refer to the second type as learning-by-doing and to the last type as learning-by-using. In our model, the final goods sector gains experience in using capital goods.
4. An advantage of this assumption is that the coexistence of different vintages cannot, by definition, be explained on the basis of incomplete depreciation of the existing capital stock as is common in 'traditional' vintage models.
5. Since the model is defined in continuous time we cannot speak of t in terms of years, although the interpretation of the model is in terms of vintages and technologies that come available in different years.
6. A notable exception is Soete and Turner (1984) who model technology diffusion in a macroeconomic model.
7. See Solow (1987) for a survey and critical discussion of the capital–energy complementarity debate.
8. Note that we basically describe the production structure in the final goods sector as a nested Cobb-Douglas function, in which capital and energy enter in one nest yielding capital services where substitution possibilities between capital and energy are assumed to be absent.
9. It is of course possible to generalize here and to model a separate sector producing the brands and selling these to the firms producing the capital. In such a setting, firms in the capital production sector would be willing to buy the patent to produce the specific brand and acquire the monopoly right to produce, provided that the profits that can be earned over time are equal to the costs of the patent (compare, for example, Grossman and Helpman 1991). We think that such a generalization, though interesting, would not add to the basic insights we want to emphasize in this chapter.
10. This budget constraint can further be omitted by virtue of Walras' law.
11. Figure 3.3 is based on a discretized version of the model with the following parametrization: $\alpha = 0.5$, w = 1 (numeraire), $g = 0.05$, $A_0 = 1$, $\gamma = 0$, $L = 300$, $L_f = 2$, $\psi = 1$, $p_E = 2$. The elasticity of substitution is equal to $\varepsilon = 4.2$ in the high-complementarity case and $\varepsilon = 7.4$ in the low-complementarity case. This results in $T = 11$ and $T = 6$, respectively. Note that these values for ε do not suggest any form of calibration but are chosen such that we arrive at a discrete number of vintages.
12. The generality of this result has not yet been proven, but extensive numerical experiments of these partial mechanisms have shown that the result is valid for a wide range of parameter values. Details are available upon request.
13. Figure 3.5 is based on a discretized version of the model with the following parametrization: $\alpha = 0.5$, $w = 1$ (numeraire), $g = 0.05$, $A_0 = 1$, $\gamma = 1.25$, $a = 0.2$, $\lambda = 0.5$, $L = 300$, $L_f = 2$, $\psi = 1$, $p_E = 2$. The elasticity of substitution is equal to $\varepsilon = 3.95$ in the high-complementarity case and $\varepsilon = 5.8$ in the low-complementarity case. This results in $T = 12$ and $T = 8$ respectively.
14. Figure 3.6 is based on a discretized version of the model with the following parametrization: $\alpha = 0.5$, $w = 1$ (numeraire), $g = 0.05$, $A_0 = 1$, $\gamma = 0$, $L = 300$, $L_f = 2$, $\varepsilon = 4.4$. The energy–capital ratio is equal to $\psi = 0.2$ in the low energy intensitiy case and $\psi = 4$•in the high energy-intensity case. This results in $T = 9$ and $T = 10$, respectively.
15. Note that it is not relevant whether the arrival of a new technology is unexpected or not, since we are not dealing with an intertemporal decision problem as we assumed capital to be non-durable.

4. Subsidizing the adoption of energy-saving technologies

4.1 INTRODUCTION

Governments of most industrialized countries have committed themselves to substantially reducing energy-related greenhouse gas emissions within a specified time horizon. Since improving energy efficiency is considered to be an important option to reach that goal, important questions arise concerning how and when to stimulate the adoption of energy-saving technologies (Grubb 1997, OECD 1999). Answering these questions requires insight into processes of technological change. The challenge for governments is therefore to develop policies aiming to achieve the macroeconomic or generic goals imposed by emission reduction targets, while taking into account the microeconomic characteristics that determine adoption behaviour of individual firms. With respect to technology adoption, two major stylized facts are (1) the (at least partly) irreversible nature of adoption decisions[1] which, in combination with uncertain technological change, gives an incentive to postpone adoption to limit the likelihood of regret (see for example Dixit and Pindyck 1994, Farzin et al. 1998 and Pindyck 1991); and (2) heterogeneity among firms or industries which results in the typical S-shaped diffusion patterns that we observe in reality (for example, Davies 1979, Jaffe and Stavins 1994, Stoneman 2001).

In this chapter we combine these issues and facts in an investment decision model that is characterized by uncertainty with respect to the timing and quality of new technologies, as well as by the existence of learning effects. We analyse the impact of uncertain technological progress by explicitly taking into account the option value of postponing the adoption (that is, the opportunity cost of immediate adoption). In addition we analyse the impact of uncertainty, learning and maturation of technologies on the effectiveness of investment subsidies aiming to stimulate the adoption of energy-saving technologies. We find that standard Net Present Value (NPV) analyses result in an incorrect assessment of energy savings, in both the short and the long term. The reason is that under the assumption of stochastic rates of technological progress, the energy savings that are not achieved in the short run due to the postponement of the adoption may be more than compensated for in the longer run due to the adoption of a superior technology. This implies that granting subsidies that tend to speed up adoption may have an adverse impact on long-run energy savings. We analyse the sensitivity of these results to the nature of technological progress by allowing for different assumptions with respect to learning or spillover effects, the success of innovation, the speed of quality improvement, the discount rate and the existence of physical upper bounds on a technology's energy-saving potential.

The results are interesting from a policy point of view because of the politically imposed time constraints for emission reductions. Traditional investment theories (see for example Stoneman and David 1984) would suggest the presence of a 'double dividend' associated with subsidizing the adoption of energy-saving technologies that are subject to learning effects. Not only do subsidies induce immediate adoption to meet politically imposed targets, they also induce further technological progress since technology adoption induces the 'take-off' of learning effects. In this chapter we illustrate with a simple model that investment subsidies may give rise to a trade-off between early adoption of relatively inferior technologies on the one hand and late adoption of relatively superior technologies on the other hand. These results translate directly into a trade-off between short- and long-run emission reduction targets. The implications are exacerbated once uncertainty is recognized as an important investment decision parameter.

There are several related articles on investment under uncertainty, in which learning plays an important role. However the analysis in this chapter differs from that literature in that we analyse the effect of investment subsidies on energy-saving technology diffusion in terms of resulting aggregate energy savings. In doing so, we focus on the uncertainty of technological progress rather than, for example, output price uncertainty (for example Majd and Pindyck 1989), investment cost uncertainty (Purvis et al. 1995), ecological uncertainty or uncertainty with respect to the future costs and benefits of reducing environmental damage (Pindyck 2000, 2002). Similar to Balcer and Lippman (1984), technology adoption in our model occurs if the technology lag exceeds a certain threshold, but we explicitly model a learning or spillover effect. Apart from learning, we do not consider other types of firm interactions, such as network externalities and strategic behaviour (cf. Choi 1994, Hoppe 2000, Moretto 2000). We do however specifically acknowledge firm heterogeneity (cf. Alvarez and Amman 1999). Finally, we ignore explicit non-convexities in environmental damages that necessitate adoption (as done in Dosi and Moretto 1997).

The set-up of this chapter is as follows. In section 4.2 we develop a simple model that captures the essence of investment under technological uncertainty. In section 4.3, we analyse the consequences of uncertainty for optimal investment behaviour in terms of cumulative energy savings. To this aim we derive optimal adoption behaviour, first ignoring and then incorporating the option value of postponement. Section 4.4 assesses the effects of subsidy schemes on the diffusion of technologies. We analyse the sensitivity of the model in terms of cumulative energy savings by allowing for different assumptions with respect to the nature of technological progress. Section 4.5 concludes.

4.2 THE MODEL

In order to analyse the effects of uncertainty, technological progress and subsidies on adoption behaviour, we model a simple economy with $\overline{N}$ firms that can potentially benefit from the adoption of a specific technology in terms of reducing the amount of energy used. Over time, better vintages (indexed *i*) of that technology become available (with identical purchase price *K*).[2] To model irreversibility, we simply assume that firms can invest only once (see Farzin et al. 1998 for a more general model allowing for multiple investments). This is clearly an extreme case of an irreversible investment, but the qualitative results of the model also apply to other cases where investments are at best partially reversible (for example, when there are scrap markets for obsolete technologies), or when firms are allowed to invest more than once.

Vintage *i* can provide a maximum rate of energy savings (measured in monetary terms) equal to R_i. Following microeconomic evidence, we assume that firms (indexed $n = 1, \ldots, \overline{N}$) are able to reap only a fraction (between zero and one) of the maximum potential energy savings, and that the fraction is firm-specific (that is, $0 < \theta_n \leq 1$). In addition, we assume that firms can be ranked according to that parameter $\theta_n(n)$ with $\partial\theta_n / \partial n < 0$.[3] In other words, all else being equal, firm number 1 ($\overline{N}$) is most (least) likely to adopt the technology as its per-period monetary savings from the adoption of vintage *i* are equal to R_i (close to zero).

In the simulations presented in this chapter (see sections 4.3 and 4.4), we adopt the following specification for the benefits of adoption that can be captured by firm *n*:

$$\theta_n = \frac{1}{2} - 4\left(\frac{2n - \overline{N}}{2\overline{N}}\right)^3. \tag{4.1}$$

This specification of the benefit parameter θ_n implies a non-uniform distribution of benefits from adoption among firms. The distribution is chosen such that the majority of the firms is located around the middle of the distribution space. These firms are characterized by an 'average' benefit from adoption. A minority of the firms are located at both ends of the distribution space. Figure 4.1 illustrates this: firms with a low index (n) benefit relatively highly from adoption (and vice versa), while the majority of firms are clustered around the 'average' benefit from adoption.

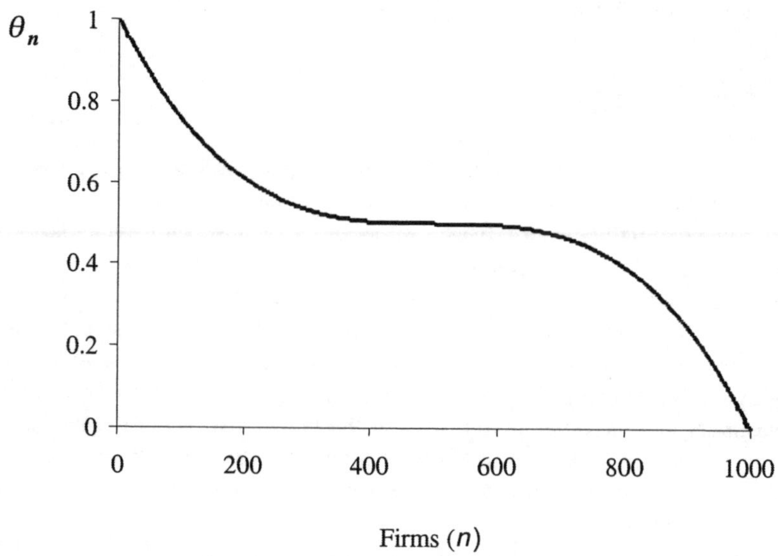

Figure 4.1 *Benefits of adoption for firm n ($n = 1, \ldots, \overline{N} = 1000$).*

The technology has an infinite lifetime and its performance, once adopted, does not change over time. Therefore there is no uncertainty with respect to a new vintage's performance when it is adopted, and hence the discounted value of the instantaneous profit stream from adopting vintage i for firm n can be defined as:

$$V_n(R_i) = \int_{t=0}^{\infty} \theta_n R_i e^{-rt} dt = \frac{\theta_n R_i}{r}, \tag{4.2}$$

where r is the (exogenous) discount rate.

Following Farzin et al. (1998) we assume that in each short period *dt* there is a given probability that a new improved vintage of the technology is discovered. Assuming a jump process, let the likelihood of a discovery in that very short period be denoted by $\lambda_t dt$, and the actual size of the jump by u_t. Then technological improvement can be modelled as follows (we omit vintage-indices if possible):

$$dR_t = \begin{cases} u_t \textit{ with probability } \lambda_t dt \\ 0 \textit{ with probability } 1-\lambda_t dt \end{cases}. \tag{4.3}$$

To model endogenous technological progress through external learning-by-doing, we assume that the likelihood λ_t of a new improved variety being discovered consists of an exogenous part (λ_0) and a part that is an increasing function of the number of firms that have already adopted the technology (N_t):

$$\lambda_t = \lambda(N_t, \lambda_0), \tag{4.4}$$

with $\partial\lambda/\partial N > 0$ and $\partial^2\lambda/\partial N^2 < 0$. This formulation captures the idea that the probability of the arrival of an improved version of the technology increases with the number of firms that is using it, representing a learning effect that is external to each individual adopter.[4] The underlying mechanism may be via the producer being informed through feedback from early adopters. Due to our assumption that firms invest only once, only those firms that have not yet adopted the new technology can reap the returns of improvements of technological performance: early adopters generate a positive externality to all other firms that are considering to purchase the technology (Kapur 1995). Essentially this reinforces the irreversible nature of adoption decisions for firms and emphasizes the existence of lock-in effects.[5] Furthermore we assume that external learning-by-doing occurs at a decreasing rate due to the increased probability of duplication.

In addition to uncertainty associated with the timing of the emergence of new, improved vintages, we assume that the improvement itself is also *ex ante* uncertain. We formalize this by assuming uncertainty with respect to the size of the jump in energy efficiency. Furthermore we assume that there is a maximum feasible efficiency for the technology (denoted by $\overline{R}$), and that the actual improvement of the technology in terms of new vintages is a decreasing function of the quality of the technology already attained. In other words we assume that a technology matures over time. Using $\overline{u}_t$ to represent the maximum possible increase in the quality of the technology at time t, we assume

$$\bar{u}_t = u\left(R_t, \bar{R}\right) \tag{4.5}$$

with $\partial u / \partial R < 0$. We assume that the realized technological improvements are uniformly distributed within the range $[0, \bar{u}_t]$, and hence the associated density function can be defined as follows:

$$f(u_t) = \begin{cases} \dfrac{1}{\bar{u}_t} & \text{for } 0 < u_t \leq \bar{u}_t, \\ 0 & \text{otherwise.} \end{cases} \tag{4.6}$$

The firm's decision process can now be summarized as follows. Until the investment is actually made, the firm has to decide whether it is going to invest or postpone the decision to invest. When exercising the option to adopt, the firm gets the pay-off $V_n(R_i)$ (that is, the present value of the profit stream as defined in equation 4.2) and pays the net adoption cost which equals the investment cost (K) minus the amount of subsidy received (S). Using equation (4.2), firm n's net present value of the decision to adopt vintage i, $\Omega_n(R_i)$, is given by:

$$\Omega_n(R_i) = V_n(R_i) - (K - S) = \frac{\theta_n R_i}{r} - (K - S). \tag{4.7}$$

This value is often referred to as the 'termination value' of the option to adopt. If the investment is perfectly reversible and firms can invest as often as they like, investment will take place whenever the termination value is positive (and hence the simple NPV rule applies). If investments are however not perfectly (and costlessly) reversible, having adopted a particular vintage can give rise to regret in future periods if new vintages arrive that outperform the purchased version. Therefore the firm has to take into account the option to postpone the investment decision into the future.[6] The value of postponing the investment for a very short period dt equals the difference between the maximum discounted value of the investment option as evaluated in that next period, and the discounted value of investing today (the termination value as given by equation 4.7).

The former includes the expected capital gain in the form of the arrival of a new improved version of the technology, and is usually referred to as the 'continuation value', which for the nth firm can be written as follows:

$$F_n(R) = \frac{1}{1+rdt} E[F_n(R+dR)]. \tag{4.8}$$

The difference between equation (4.8) and equation (4.7) is the option value of postponing the investment. In each short period *dt*, the firm compares the termination value of the option to adopt equation (4.7) with continuation value equation (4.8). The opportunity costs of postponing adoption at time *t* are foregone profits associated with the fact that the best available technology at that time is not implemented. They increase over time because of ongoing technological progress, and hence the termination value will dominate the continuation value at some future point in time. Hence for each firm there exists a critical technological quality or savings potential, denoted R_n^*, at which the firm is indifferent between investing and postponing. Adoption occurs as soon as the actual savings exceed that critical value.

To illustrate the basic mechanisms of the model, we turn in the next section to a simplified version of the model that highlights the importance of accounting for technological uncertainty. The implications of the learning-by-doing effects and the technological maturation are postponed until section 4.4.

4.3 THE IMPACT OF UNCERTAINTY ON TECHNOLOGY ADOPTION

To illustrate the impact of stochasticity with respect to the rate of technological progress on adoption behaviour, we simplify the model described in the previous section by assuming the rate of technological progress to be exogenous (that is, there is no learning-by-doing), and by assuming that there is no physical limit with respect to the energy-saving benefits (that is, there is no maturation). Technically we specify equations (4.4) and (4.5) as $\lambda_t = \lambda_0$ and, $\bar{u}_t = \bar{u}$ respectively. Hence the process of technological progress defined in equation (4.3) can be written as:

$$dR = \begin{cases} u \textit{ with probability } \lambda_0 dt \\ 0 \textit{ with probability } 1-\lambda_0 dt \end{cases} \tag{4.3'}$$

and the associated density function (4.6) as:

$$f(u)=\begin{cases}\dfrac{1}{\overline{u}} & \text{for } 0\le u\le \overline{u},\\ 0 & \text{otherwise.}\end{cases} \tag{4.6'}$$

Equation (4.6') implies that the expected technological jump equals $\overline{u}/2$.

To solve the model, we follow Farzin et al. (1998) and determine the critical quality of the technology (R_n^*) by equating the termination and the continuation value (equations (4.7) and (4.8), respectively); see Appendix 4.A for technical details. This critical quality of the technology at which firm n will exercise its option to adopt equals:

$$R_n^* = \frac{r(K-S)}{\theta_n} + \frac{1}{2}\frac{\lambda\overline{u}}{r}. \tag{4.9}$$

This expression reveals that the critical technology level in the presence of uncertainty is equal to the critical technology level in the theoretical case that firms simply apply the standard Net Present Value rule (the first term on the RHS of equation (4.9)),[7] plus a factor that is associated with uncertainty (the second term on the RHS). Technologies that are profitable from a NPV perspective are thus not necessarily profitable when taking uncertainty into account. Evidently the higher the expected capital gains (either as a result of a higher likelihood of technological progress λ_0, or because of larger expected improvements ($\overline{u}/2$)), the better the critical performance of the technology should be in order to trigger adoption.

In the absence of learning effects, we can derive an analytical solution for the number of firms that has adopted at a particular point in time as a function of the quality of the technology at that time (R_t). When firms are assumed to apply the NPV rule (see note 6), the number of firms that has adopted at time t can be determined by combining equation (4.1) and R_n^{NPV}. This yields:

$$N_t^{NPV} = \frac{\overline{N}}{2} + \overline{N}\sqrt[3]{0.25\left(0.5 - \frac{r(K-S)}{R_t}\right)}. \tag{4.10}$$

When investments are irreversible and firms take uncertainty into account, the number of adopters at time t can be found by combining equations (4.1) and (4.9):

$$N_t^{UNC} = \frac{\overline{N}}{2} + \overline{N}\sqrt[3]{0.25\left(0.5 - \frac{r(K-S)}{R_t - 0.5\lambda\overline{u}/r}\right)}. \quad (4.11)$$

For illustrational purposes, we have taken a set of parameter values and conducted a Monte Carlo simulation with 1000 runs based on random draws from the distributions of the uncertain parameters u and λ.[8] Figure 4.2 depicts the diffusion path of the technology, averaged over the 1000 simulation runs. It illustrates that adoption lags behind when firms take into account uncertainty about technological progress in their investment decisions.

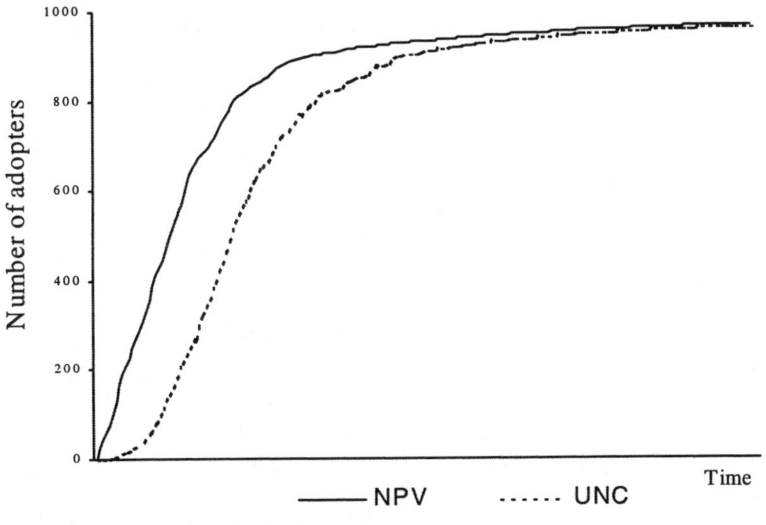

Figure 4.2 Diffusion paths of the technology in the case of investment under uncertainty (UNC) and investment under the standard Net Present Value rule (NPV).

The consequences of these differences in adoption behaviour for aggregate energy savings are illustrated in Figures 4.3a–c. The figures depict, in a discrete-time version of the model, the per-period energy savings (measured in monetary terms) that are achieved under the two investment rules for three different levels of technological sophistication.

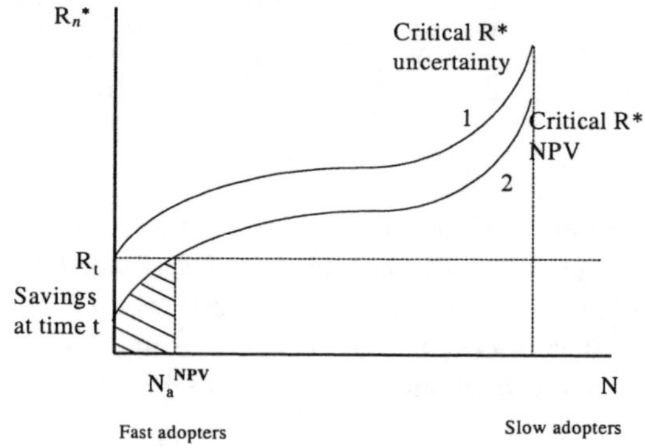

Figure 4.3a

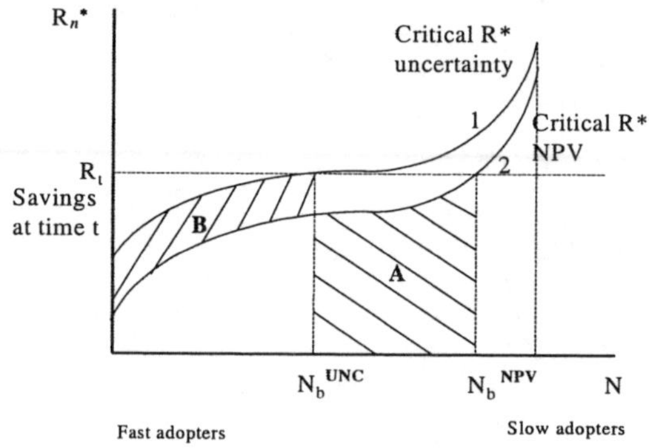

Figure 4.3b

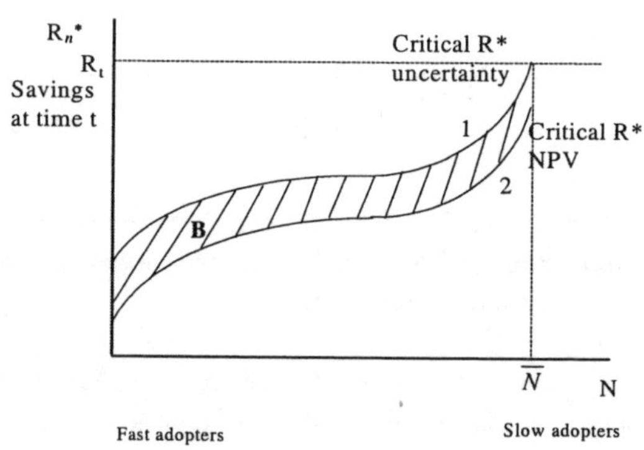

Figure 4.3c

Figure 4.3 *Per-period energy savings for different investment rules: (a) infant technology, adoption only under NPV rule; (b) technology has progressed, different levels of adoption under NPV and uncertainty rule; and (c) mature technology, all firms have adopted*

The upward-sloping curves depict the critical savings level (on the vertical axis) that triggers the investment for each firm type n (ranked on the horizontal axis according to decreasing net benefit from adoption). As is clear from equation (4.9), R_n^* is an increasing function of n (as θ_n is decreasing in n), and the critical value taking into account uncertainty strictly exceeds the critical value based on the NPV criterion alone (the difference being ($\lambda\bar{u}/2r$)). Therefore as technology progresses (R increases along the vertical axis), one by one firms will adopt, depending on the fraction of the benefits they can reap.

In Figure 4.3a, the technology is still in its infancy; the amount of energy saved by the technology (R_t) is fairly low. The level of technological quality is not sufficient to trigger investment by adopters that take uncertainty into account, and their total savings (at one period in time) are thus zero in this instance. The number of firms that invest when they behave according to the NPV rule is positive and equals N_a^{NPV}. The per-period savings are then equal to the shaded area under curve 2.

In Figure 4.3b, the technology has improved somewhat further. In the case that firms take uncertainty into account, the number of adopters equals N_b^{UNC}, whereas the number of adopters when firms behave according to a NPV rule is N_b^{NPV}. The difference in savings between the two 'regimes' can be calculated by subtracting surface B from surface A in Figure 4.3b. The net result denotes the additional savings when firms invest according to a NPV rule as compared to investment under uncertainty. At initial stages of technological progress (where energy savings are fairly low), this difference is positive. However when the technology improves further, the difference becomes negative.

The extreme of this final stage is depicted in Figure 4.3c, in which the technology has progressed to the extent that all firms have found it profitable to adopt the technology under consideration, independent of the adoption criterion used. In that case, the total savings in the regime where firms invest taking uncertainty into account unambiguously exceed the savings in the regime where firms invest according to a NPV rule. This is due to the fact that although the firms that take uncertainty into account adopt at a later point in time, they also adopt relatively better technologies.

The fact that adoption lags behind when firms take uncertainty into account over technological progress in their investment decisions, is thus not necessarily bad from an energy-saving point of view. In the longer run, total energy savings will increase due to the adoption of improved technologies. In other words, neglecting the effects of uncertainty on the adoption behaviour of firms results in an overestimation of short-run energy savings, while long-run energy savings tend to be underestimated.

4.4 THE EFFECTIVENESS OF SUBSIDIES

So far we have ignored the role of learning effects and the possibility of technological maturation. In this section we relax these assumptions and analyse the impact of adoption subsidies on (1) the timing of adoption, and (2) aggregate energy savings over time while incorporating learning and maturation into the model.[9] For tractability, we numerically analyse these effects by comparing the results for a low and a high subsidy level.

To be able to perform a numerical analysis, we need to specify the probability of the arrival of a new vintage and the size of the jump as described in equations (4.4) and (4.5). The likelihood of a technological improvement (λ_t) is specified as:

$$\lambda_t = \lambda_0 + \alpha N_t^{\beta} . \tag{4.4'}$$

This formulation captures the idea that the probability of technological improvement increases with the number of firms using that technology. This represents the learning effect. We assume that learning occurs at a decreasing rate due to the increased probability of duplication ($\beta < 1$). The maximum size of the improvement of the technology is specified as:

$$\overline{u}_t = \overline{R} - \gamma R_t . \tag{4.5'}$$

The technology has thus matured when $R_t = \overline{R}/\gamma$. At that level the technology no longer improves and uncertainty regarding technological progress is completely absent.

Totally differentiating the number of adopters at each point in time (N_t) as defined in equation (4.11) with respect to the subsidies, we find that subsidies unambiguously speed up the diffusion process.[10] The effect of subsidies on the level of aggregate energy savings is however ambiguous. Although subsidies speed up adoption and thus increase short-run energy savings, accumulated energy savings can be negatively affected since earlier adoption also implies adoption of relatively inferior technologies. Since we assume that firms invest only once, high subsidies may thus contribute to the occurrence of a lock-in that is undesirable from a policy perspective. Putting the argument the other way around: a low subsidy level leads, *ceteris paribus*, to a relatively slow diffusion of energy-saving technologies, but therefore also to the adoption of relatively better technologies. As a result, accumulated energy savings can be positively affected by low subsidies in the longer run since low subsidies can avoid a lock-in into inferior technologies.

In order to illustrate this, we again conducted a series of Monte Carlo simulations with 1000 runs based on random draws from the distributions of the parameters u and λ. We analysed aggregate energy savings over time for two different subsidy levels: a low and a high one. To determine the robustness of the results for specific parameter values, we conducted an extensive sensitivity analysis with respect to α (the degree of learning), r (the discount rate), γ (the speed of technology maturation), λ_0 (the exogenous arrival of new technologies; innovation success) and $\overline{R}$ (the level of maturation).

Depending on the assumptions regarding technological maturation and learning, we can distinguish four versions of the model. Each version is characterized by a specific pattern of expected technological improvement. The four possibilities are depicted in Figure 4.4.[11]

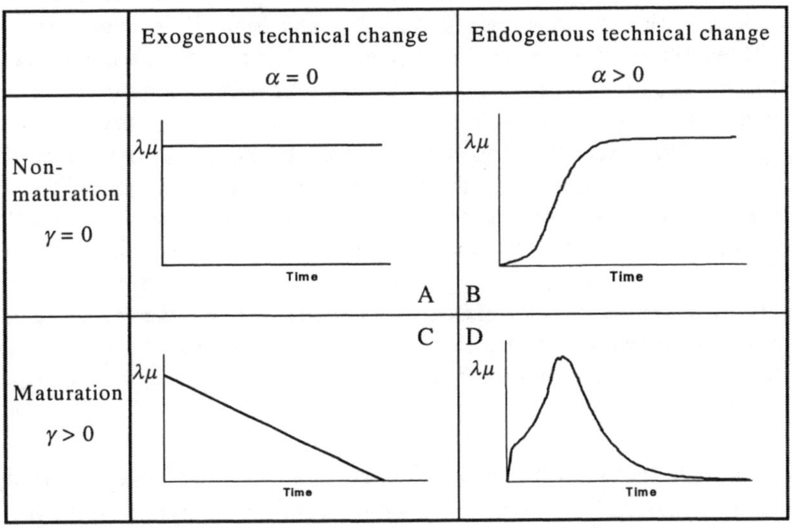

Figure 4.4 Expected technology improvement under four different assumptions with respect to the parameter values

Quadrant A reflects the simple version of the model in which maturation and learning are neglected. One can think here of a small open economy in which innovations take place abroad, and are thus exogenous to the domestic country, whereas domestically no knowledge spillovers occur. In this case the expected technological improvement is constant over time since both λ and u are exogenous and constant. This is the case that was discussed in section 4.3.

In quadrant B learning is introduced, while technological improvement is still unbounded. This implies that u_t is constant, while λ_t is a positive function of the number of adopters (see equation 4.3'). The technology is expected to improve at an ever faster rate so long as the number of adopters increases, but as soon as the model converges to the steady state, expected technological improvement becomes constant. The steady state is characterized by a situation in which all firms have adopted the technology ($N_t = \overline{N}$).

Quadrant C depicts the pattern of technological improvement in the absence of learning, but with bounded technological improvement. Now, λ_t is constant while u_t is a decreasing function of the quality of the technology (R_t) already attained (see equation 4.5'). A steady state will be reached with a constant number of adopters that is potentially smaller than $\overline{N}$, a constant technological quality, and no uncertainty regarding technological progress.

Finally, the hump-shaped pattern in Quadrant D results from the combination of learning and maturation. An increase in the number of adopters results in an increasing probability λ_t of a technology improvement. On the other hand, the resulting increase in the quality of the technology implies a decline in the remaining potential for further improvement. For reasonable parameter values, the second effect starts to dominate the first effect after a while, leading to the hump-shaped curve. A steady state will be reached that is qualitatively similar to the steady state described in Quadrant C.

We will now discuss the effects of subsidies on aggregate energy savings for the different cases that can be distinguished. Extensive sensitivity analysis for different parameter values has shown two possible outcomes of the model:

1. A high subsidy leads to a relatively high level of aggregate energy savings in both the short and the long run. In this case the effect of the higher number of adopters caused by the higher subsidy (which will be referred to as the Scale Effect) dominates the low average quality of the adopted technologies associated with higher subsidies (further denoted as the Quality Effect). In other words, subsidies are effective in stimulating energy saving because they trigger large-scale adoption of energy-saving technologies, in spite of the fact that this implies a lock-in into relatively inferior technologies. This case is illustrated in Figure 4.5.
2. A high subsidy leads to a relatively high level of aggregate energy savings in the short run, but to a relatively low level in the long run, since the slow adoption under a low subsidy regime is compensated by the adoption of relatively better technologies in the long run. So in this case the Scale and Quality Effects compete with each other, which

results in a trade-off between the short and the long run. In other words, the lock-in effect causes the investment subsidies to be counter-effective in the long run; see Figure 4.6.

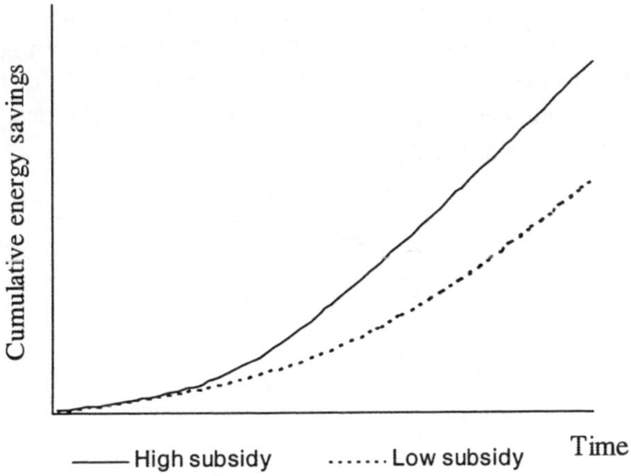

Figure 4.5 The Scale Effect

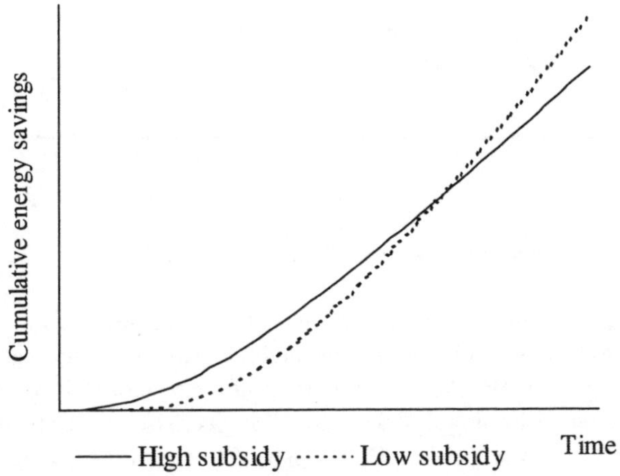

Figure 4.6 Trade-off between short- and long-run energy savings due to the opposed impact of the Scale and Quality Effects

The Scale Effect can only dominate in the long run if the technology matures over time. In the absence of maturation, all firms eventually adopt the technology since the unbounded technological improvement assures that in the end even for the last firm – which has lowest benefit from adoption (see Figure 4.1) – the quality of the technology exceeds its critical level as defined in equation (4.9). When we allow for technological improvement to be bounded from above, at the time of maturation of the technology a stationary situation can arise in which not all firms have adopted the energy-saving technology (see Figure 4.7).

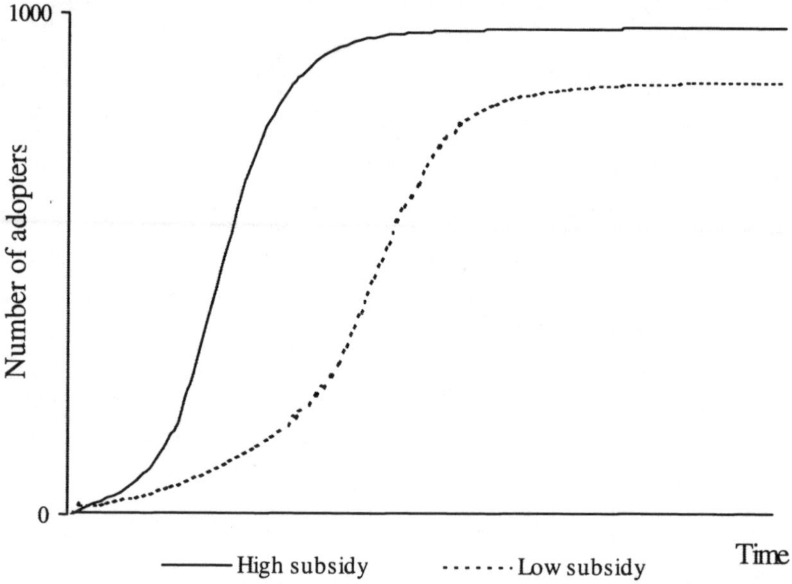

Figure 4.7 Number of adopters if technology matures

The reason is that for those firms with low benefits from adoption, the quality of the technology does not exceed the critical quality that they face. A high subsidy scheme then results in relatively more firms adopting the energy-saving technology, since subsidies decrease the critical technology level (see equation (4.9)). This Scale Effect can outweigh the effect associated with the adoption of relatively inferior technologies. The series of Monte Carlo simulations we performed indicates that this condition is met if the diffusion process is relatively slow in the absence of subsidies. This is intuitively clear as slow diffusion reduces the chance that the Quality Effect starts to dominate. Diffusion is relatively slow if we have, *ceteris paribus*, a low learning rate (α), a high speed of exogenous arrival of new technologies (λ_0), a high level of exogenous maturation ($\overline{R}$) a high discount rate (r) or a fast speed of maturation (γ). To see this, recall that equation (4.9) implies that an increase in expected technology improvement ($\lambda\overline{u}$) or discount rate (r) will raise the critical technology level. This slows down adoption since the quality of the technology should be higher to trigger adoption. Intuitively, when large technological improvements are to be expected, firms postpone adoption. The same arguments hold for a high speed of technology maturation (γ): large jumps in technological improvement lead to a delay in adoption since better technologies can be expected within a short period of time. A low learning rate implies that late adopters reap only limited benefits from early adopters' experience, as a result of which technological improvement is relatively slow.

In sum, if (for various reasons) diffusion of new technologies is relatively slow, then subsidies can play an important role in stimulating adoption, and the resulting increase in the number of adopters can be sufficiently large to compensate for the relatively inferior technologies adopted. On the contrary, if diffusion is relatively fast in the absence of subsidies, the number of additional adopters due to a higher subsidy is limited and is not necessarily sufficient to outweigh the negative effect of adopting relatively inferior technologies. This leads to the dominance of the Quality Effect: increasing subsidies do have a beneficial effect on energy savings in the short run, but a negative effect in the long run due to the lock-in in technologies of a relatively inferior quality. We summarize these results in Figure 4.8.

	Exogenous technical change $\alpha = 0$	Endogenous technical change $\alpha > 0$
Non-maturation $\gamma = 0$	*Quality Effect* A	*Quality Effect* B
Maturation $\gamma > 0$	C *Scale Effect* *Quality Effect*	D *Scale Effect* *Quality Effect*

Figure 4.8 Effect of subsidies on aggregate energy savings under four different assumptions

These results are in contrast with traditional investment theories (see for example Stoneman and David 1984) suggesting that the existence of learning effects enhances the effectiveness of investment subsidies, since they not only induce immediate adoption in itself but in doing so also stimulate further technological progress by triggering learning effects. However in general the quality of existing and new technologies is not fixed: often their performance improves over time while the speed and extent of the improvements is uncertain. We found that because of this, investment subsidies may be counter-effective in the long run, since they can induce investments in relatively inferior technologies and thereby diminish the market potential for improved versions of a technology.

4.5 CONCLUSIONS

Governments of many industrialized countries have committed themselves to achieving a certain level of greenhouse gas emission reduction within a certain short period of time. Subsidizing the adoption of energy-saving technologies is one instrument used to reach this goal. In this chapter we developed a model integrating insights on technology adoption and investment behaviour under uncertainty in order to analyse the effects of uncertainty, the nature of technological progress and subsidies on adoption behaviour in both the short and the long run. A first result that we derived is that neglecting the effects of uncertainty on investment behaviour results in an overestimation of short-run savings and an underestimation of long-run savings. Subsidies were shown to raise the speed at which firms adopt. This unambiguously fosters energy savings in the short run. In the longer run however, account has to be taken of the quality of the technologies that are being adopted. The delayed response of firms receiving a low subsidy results in – on average – the adoption of better technologies. In the long run, this quality effect may dominate the short-run effect of increased adoption of technologies in terms of the level of aggregate energy savings. In other words, increasing investment subsidies for energy-saving technologies can be counterproductive from a policy perspective as they may favour a lock-in into relatively inferior technologies.

In sum, we find that investment subsidies are most effective if they stimulate aggregate energy savings while avoiding a lock-in into inferior technologies. In the context of our simple model, this criterion is most likely to be fulfilled if the diffusion process is slow in the absence of subsidies. In this situation, subsidies significantly increase the number of adopters and thereby aggregate energy savings, while a lock-in into inferior technologies is likely to be avoided. The reasons for a slow rate of adoption can be various: a low degree of knowledge spillovers, a high discount rate, or the expectation of significant technological improvements in the near future.

The politically imposed time constraints for realizing greenhouse gas emission reductions have induced policy-makers to increase investment subsidies to stimulate the adoption of energy-saving technologies. The answer to the question whether this is a beneficial strategy in terms of accumulated energy savings depends crucially on the endogenous nature of technological progress and the degree of uncertainty.

APPENDIX 4.A

To solve the model, we follow Farzin et al. (1998) and calculate the continuation value (4.7). The expected increase in the value of the option to adopt is a function of the exogenous rate of technological progress ($\bar{\lambda}$) and the maximum jump size ($\bar{u}$). These parameters are exogenous to individual firms. If a new vintage of the technology becomes available, the question is whether it will trigger adoption.

Clearly the probability of adoption depends on both the critical value of energy savings that triggers adoption (R_n^*, which needs to be determined) and the current state of the technology in terms of energy savings, R. The improvement will trigger adoption if the jump in the quality of the technology is sufficiently large ($R_n^*-R \leq u \leq \bar{u}$), whereas no adoption will take place if the jump in quality is relatively small ($0 \leq u \leq R_n^*-R$). The probabilities of these cases occurring can be determined using equations (4.3') and (4.6'). This yields:

$$E[F_n(R+dR)]-F_n(R) = \tag{4.A.1}$$

$$\lambda dt \left\{ \int_{u=0}^{R_n^*-R} \frac{1}{\bar{u}} F_n(R+u)du + \int_{u=R_n^*-R}^{\bar{u}} \frac{1}{\bar{u}} (\Omega_n(R+u))du - F_n(R) \right\}.$$

The first term between brackets on the right-hand side (RHS) of (4.A.1) reflects the expected value of continuation if the jump in the quality of technology is relatively small. If this jump (which occurs with likelihood λdt) is relatively small, the technology parameter will not exceed the critical value R_n^* after the jump, the investment option is kept alive and the decision to invest is postponed. The second term between brackets on the RHS reflects the expected termination value: if the improvement in the technology is relatively large, the option to invest is executed. Combining equations

(4.A.1) and (4.8) yields the value of the option to adopt as a function of the quality of the technology under consideration:

$$F_n(R) = \frac{\lambda}{r+\lambda}\left[\int_{u=0}^{R_n^*-R} \frac{1}{\bar{u}} F_n(R+u)du + \int_{u=R_n^*-R}^{\bar{u}} \frac{1}{\bar{u}}(\Omega_n(R+u))du\right]. \quad (4.A.2)$$

By definition, at $R=R_n^*$, investing is optimal after the next jump. Substituting $R = R_n^*$ into equation (A2) and using (4.7), we find:

$$F_n(R_n^*) = \frac{\lambda}{r+\lambda}\left[\int_{u=0}^{\bar{u}} \frac{1}{\bar{u}}(V_n(R_n^*+u)-(K-S))du\right] = \quad (4.A.3)$$

$$\frac{\lambda}{(r+\lambda)}\left[\int_{u=0}^{\bar{u}} \frac{\theta_n}{r}(R_n^*+\frac{1}{2}\bar{u})-(K-S)\right].$$

In the optimum it must hold that at the critical technology level, the value of the adoption project for the nth firm is equal to its termination value:

$$F_n(R_n^*) = \Omega_n(R_n^*). \quad (4.A.4)$$

Combining equations (4.A.3) and (4.7), we derive equation (4.9) in the main text.

NOTES

1. Investments are said to be irreversible if not all costs associated with the technology adoption can be recouped. Two important sources of irreversibility are the installation costs, and the fact that the resale value of the machinery generally falls short of the purchase price.
2. For simplicity we assume that all new varieties of the technology require the same gross investment expenditures (K). If we assume that there is a scrap market for old technologies, K should be interpreted as the net adoption costs, that is, the costs of adopting the new technology minus the scrap value of the old one.
3. Although we assume a discrete number of firms, we characterize θ by an approach in terms of continuous time, assuming that the ranking of θ is built up of a sufficiently large number of small differences in benefits.
4. We thus ignore learning-by-using, the effect that adopters themselves gain experience with the use of the particular technology and hence realize a quality improvement and/or a cost reduction at the plant level (Rosenberg 1976).
5. Note that we do not explore the fact that adoption by one firm may hurt its rivals, for example due to increased competition in the product market (business stealing effect). As a

result, firms in our model are not involved in strategic behaviour and hence we do not explore issues of first and second-mover advantages or 'waiting games' that obviously often play an important role in technology adoption (see Hoppe 2000, Moretto 2000). However because of the learning effect spilling over from early to later adopters, our model does include a type of network externality.

6. Note that this result also holds if firms are able to invest multiple, but not an unlimited number of times; see Farzin et al. (1998).
7. In a world without irreversibility (that is, if firms can invest an infinite number of times), the appropriate decision criterion is the standard NPV rule, which says that investment should occur as soon as the expected net present value of adoption is positive. This holds if $\Omega_n(R_i)$ (equation 4.7) is larger than or equal to zero. The NPV criterion thus equals $R_n^{NPV} = r(K-S)/\theta_n$.
8. The following parameter values were used: $\overline{u} = 0.03$, $\overline{\lambda} = 0.3$, $R_0 = 0.1$, $\overline{N} = 1000$, $K = 1$, $S = 0.3$ and $r = 0.1$. We refer to these values as the baseline scenario.
9. Unlike Pindyck (2000, 2002) who deals with the issue of optimal timing (and level) to adopt an environmental policy to reduce greenhouse gases, we focus on the effectiveness of investment subsidies given the fact that the government has already adopted a policy to achieve a certain level of emission reduction within a certain time span. (In a way this difference in research question reflects the difference in position between the USA and Europe with respect to the Kyoto protocol.)
10. Note that we make the assumption here that firms take λ_t and u_t as constants at the time of their decision, since the development of λ_t and u_t according to equations (4.4') and (4.5') is a process external to individual firms; it is the aggregate outcome of the decisions of all other firms in the economy, which we assume an individual firm cannot oversee.
11. We used the baseline scenario (see note 7) and put $\alpha = 0.03$, $\beta = 0.75$ and $\gamma = 0.03$.

PART III

Empirical Analyses

5. International comparisons of sectoral energy and labour productivity: stylized facts and decomposition of trends

5.1 INTRODUCTION

Ever since Solow (1957) held his famous 'residual' responsible for most of the observed economic growth, broad consensus has existed that long-run economic growth is caused by technology-driven (total) factor productivity growth. This has led economists to focus on the role of productivity and technology in their quest to understand economic growth. The quest has not been confined to economic theorizing about growth and technological change, but includes also empirical work on the sources of economic growth. Over the last few decades, a growth accounting tradition emerged measuring the contribution of various determinants to output and productivity growth (see for example Kendrick 1961, Denison 1967, Jorgenson and Griliches 1967, Maddison 1991, Jorgenson 1995, Wagner and van Ark 1996, van Ark 1997, Barro 1997). This empirical research on productivity growth has focused almost exclusively on labour, capital and total factor productivity growth.

However at the same time increasing attention has been paid to the role of energy in production processes and economic growth. Energy is an essential factor that fuels economic growth and serves human well-being. Along with unprecedented economic growth, world primary energy use has grown enormously since the middle of the nineteenth century. The energy crisis of the 1970s, and more recently the environmental problems associated with economic growth and increasing energy use, have induced empirical research on energy productivity or energy intensity developments and its determinants (see for example Berndt 1978; Jorgenson 1984, 1986; Howarth et al. 1991; Morovic et al. 1987, 1989; Schipper and Meyers 1992; Rosenberg 1994; Miketa 2001). Moreover it made most governments in OECD countries strive explicitly for sustainable development, aiming to decouple economic growth and environmental pressure. In a more operational sense this implies that not only labour productivity but also energy productivity should increase. Against this background this chapter offers international comparisons of energy and labour productivity developments for the period 1970–97, distinguishing 13 sectors – including ten manufacturing sectors, services, transport and agriculture – and 14 OECD countries. In doing so, we build upon insights from the traditional empirical growth literature as well as from the literature on energy intensity developments.

The level of sectoral detail in our dataset allows us to trace back macroeconomic energy and labour productivity developments to developments at the level of individual sectors, in order to correct as much as possible for the impact of 'structural effects' on productivity trends. This is important since observed aggregate productivity trends are not directly attributable to technological change in individual sectors, but are also the result of changes in the distribution of production factors among sectors. The latter is due to the fact that some sectors produce more value added per unit of input (energy or labour) than others, because some activities require more capital, higher labour skills and/or technology than others. Hence understanding technology-driven productivity performance requires an assessment of productivity performance in individual sectors (see also for example Dollar and Wolff 1993, Wagner and van Ark 1996, Jorgenson 1984). In order to identify to what extent aggregate productivity trends are to be explained from, respectively, shifts in the underlying sector structure and efficiency improvements in individual sectors, we decompose per country changes in overall productivity performance into a so-called 'sectoral effect' and an 'efficiency effect'. In several respects, our decomposition differs from most other decomposition studies (see for example van Ark 1996, Unander et al. 1999), most notably by the combination of a relatively high level of sectoral detail with a wide range of countries and a simultaneous exploration of productivity performance along the two dimensions of energy and labour.

By documenting the relation between energy and labour productivity growth rates we touch upon the issue of the direction of technological change. The idea that the nature of technological progress might be factor-augmenting, depending on relative factor prices and substitution possibilities, goes back to Hicks (1932) and has received attention in the theoretical and empirical literature on technological change and factor productivity developments ever since (see for example Kennedy 1962; Binswanger 1974a, b; Acemoglu 2002; Ruttan 2001). Recently the issue has also been addressed in the context of environmental policy and energy use, examining a price- or product-standard induced bias towards energy-saving technological change (see for example Newell et al. 1999, Smulders and de Nooij 2003, Taheri and Stevenson 2002). An important hypothesis in this respect is that if all technological efforts are directed towards an increase in labour productivity, energy productivity improvements might slow down because of lack of resources devoted to increasing energy efficiency – and vice versa. In this chapter we provide some empirical evidence on the existence and development of a potential bias towards either energy- or labour productivity, which might reflect biases of technological change at the level of individual sectors.

The chapter proceeds as follows. In section 5.2 we give a brief description of the data used in this study. In section 5.3 we document several stylized facts on the levels and trends in macroeconomic energy and labour productivity performance, examining the role of the manufacturing, services, transport and agriculture sectors. In section 5.4 we take a closer look at the manufacturing sector, not only because it is an important sector within the OECD – still responsible for about 40 per cent of total final energy consumption and 25 per cent of total employment – but also because it is a very heterogeneous sector in terms of production structure. Therefore we further disaggregate the manufacturing sector into ten subsectors and examine their role in driving aggregate manufacturing energy and labour productivity performance, following the same research strategy as we will use in section 5.3. Section 5.5 concludes.

5.2 DATA

The analysis presented in this chapter is based on a newly constructed database that merges energy data from the Energy Balances as they are published by the International Energy Agency (IEA) and economic data from the International Sectoral Database (ISDB) and the Structural Analysis Database (STAN), both published by the OECD. The main idea behind the construction of this database is to establish a link between economic and energy data at a detailed sectoral level. This results in the sector classification as described in Table 5.1.

The database covers the period 1970–97 and includes the following countries: Australia (AUS), Belgium (BEL), Canada (CAN), Denmark (DNK), Finland (FIN), France (FRA), West Germany (WGR), Italy (ITA), Japan (JPN), the Netherlands (NLD), Norway (NOR), Sweden (SWE), the United Kingdom (GBR) and the United States (USA). For a detailed description of the database we refer to Mulder (2003). In the remainder of this section we briefly highlight a few important characteristics as well as limitations of our dataset.

Table 5.1 Sector classification

	Sector	Code	ISIC Rev. 2 code
1.	Food and tobacco	FOD	31
2.	Textiles and leather	TEX	32
3.	Wood and wood products	WOD	331[a]
4.	Paper, pulp and printing	PAP	34
5.	Chemicals	CHE	351+352[b]
6.	Non-metallic minerals	NMM	36
7.	Iron and steel	IAS	371
8.	Non-ferrous metals	NFM	372
9.	Machinery	MAC	381+382+383[c]
10.	Transport equipment	MTR	384
11.	Construction	CST	50
12.	Services	SRV	61+62+63+72+81+82+83+90[d]
13.	Transport	TAS	71
14.	Agriculture	AGR	10

Notes:
[a] WOD excludes furniture since the sector WOD in the IEA Energy Balances excludes furniture.
[b] CHE includes non-energetic energy consumption, i.e. using energy carriers as feedstock.
[c] MAC includes Metal products (BMA, 381), Agricultural and industrial machinery (MAI, 382) and Electrical goods (MEL, 383);
[d] SRV includes Wholesale and retail trade, restaurants and hotels (RET), Communication (COM), Finance, insurance, real estate and business services (FNI), and Community, social and personal services (SOC).

We measure energy productivity by gross value added per unit of final energy consumption and labour productivity by gross value added per worker (in full-time equivalents).[1] Value added is the net economic output of a sector, measured by the price differential between the price of output and the cost of input, and comprises compensation to employees, operating surplus, the consumption of fixed capital and the excess of indirect taxes over subsidies (OECD 1998). Following the IEA, energy use is defined as final energy consumption in kilotonnes of oil equivalence (ktoe),[2] with sectoral data excluding transformation losses. Total employment is measured in full-time equivalent number of persons, including self-employed.

The value added data have been converted to constant 1990 US dollars, using 1990 expenditure purchasing power parities (PPP) as given by the OECD. In principle the theoretically most appropriate conversion factors for productivity comparisons at the sectoral level are to be based on a comparison of output prices by industry of origin, rather than on expenditure prices (see for example van Ark and Pilat 1993). Expenditure PPPs exclude the part of output that is exported, while they include imported goods produced elsewhere; they take account of differences in trade and transport margins and indirect taxes between countries; and they do not cover intermediate products. The main problem in using the production or industry-of-origin approach however is the limited availability of producer-price based PPPs, in particular for non-manufacturing sectors (van Ark 1993).[3] Hence most studies including cross-country productivity comparisons use expenditure PPPs. Moreover for an international comparison the main issue is whether there are substantial cross-country differences with respect to the drawbacks of expenditure PPPs as outlined above. We have no *a priori* reason to presume that these cross-country differences are substantial. Therefore in this study we use expenditure PPPs, enabling a systematic cross-country analysis of energy and labour productivity performance at a high level of sectoral detail.

In general it holds for each analysis of productivity developments that the lower the level of aggregation the better, but that an adequate distinction between factor-intensive and factor-extensive sectors is even more important. In this respect it is to be noted that our chemicals sector is defined at a rather aggregated level, built up from the energy-intensive subsector of industrial chemicals (ISIC 351, including basic industrial chemicals, fertilizers, pesticides and main plastic products) and the energy-extensive subsector Other Chemical Products (ISIC 352, including paints, drugs and medicines, cosmetics and cleaning products). Until consistent and internationally comparable energy and economic data become available for a more detailed breakdown of the chemicals sector, this problem will persist and energy productivity figures for the chemicals sector should be interpreted with caution.

Unfortunately, for the USA the IEA Energy Balances provide no sectoral breakdown for the consumption of oil products and natural gas within manufacturing until 1995. Instead, these volumes are included in the non-specified industry sector (NSI), and hence they are available only at the level of the aggregate manufacturing sector. For this reason, for the different USA manufacturing subsectors, we neither include data from 1995 onwards nor do we report levels of energy consumption. We do however calculate energy productivity growth rates for the different USA manufacturing subsectors, with final energy consumption defined as the sum of only coal and electricity consumption, under the assumption that the share of the sum of oil and gas in final energy consumption is more or less constant over time. Hence the documented growth rates of energy productivity for the breakdown of the USA manufacturing sector should be interpreted with caution.

5.3 MACROECONOMIC DEVELOPMENTS

In this section we explore levels and trends in macroeconomic energy and labour productivity performance, examining the role of the manufacturing, services, transport and agriculture sectors. We start with a brief overview of their sectoral shares in macroeconomic energy consumption, employment and GDP. The manufacturing sector used to be the most important sector from an energy point of view, accounting for more than 40 per cent of the world's energy use (Schipper and Meyers 1992). In the OECD the transport sector is nowadays at least as important as manufacturing in terms of energy consumption. For the sum of the 14 OECD countries included in this study, the share of total final energy consumption in transport accounted for 42 per cent in 1990, closely followed by manufacturing with 40 per cent, while services accounted for 15 per cent, agriculture for 2 per cent and construction for 1 per cent (see Figure 5.1).

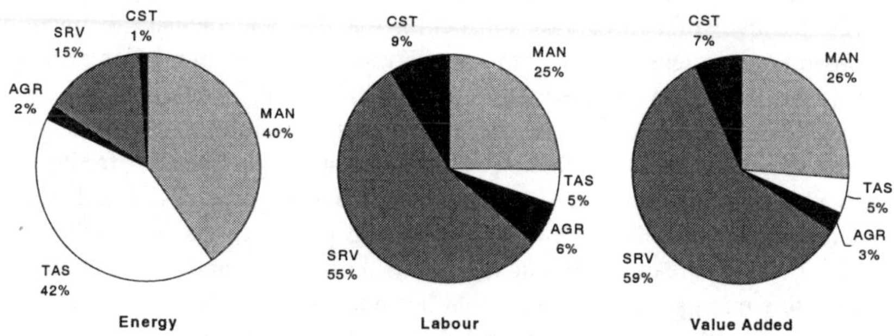

Figure 5.1 Percentage shares of non-residential final energy consumption, labour and value added by sector in 1990. Sample of 14 OECD countries.

In Figure 5.1 we compare those shares with the sector shares of total employment and value added. Our data confirm the well-known fact that for industrialized countries the highest share of total employment and value added can be found in the service sector (55–60 per cent), followed by manufacturing (25 per cent), while transport, agriculture and construction are responsible for the remaining 15–20 per cent. These shares are more or less similar for all of the 14 OECD countries included in this study (see Table 5.A.1 in Appendix 5.A).

In sum, the service sector plays a major role in terms of value added and total employment, while most energy is consumed in transport and manufacturing. For transport in particular there is a large contrast between the share of total energy consumption on the one hand and the share of total value added and employment on the other hand.

Within the OECD, the absolute level of energy consumption and employment grew over the last decades, but so did economic activity. In this chapter we take this volume effect into account by using energy and labour productivity as indicators to relate, respectively, final energy consumption and employment to the level of economic activity.[4] In the remaining part of this section we provide a cross-country comparison of energy and labour productivity levels, followed by a cross-country comparison of energy and labour productivity growth rates.

Comparing Energy and Labour Productivity Levels

To compare cross-country energy and labour productivity performance at the macroeconomic level, we calculated for each country the energy and labour productivity levels for the sum of the manufacturing, transport, services and agriculture sectors.[5] In Figure 5.2 we plot the development of these macroeconomic energy and labour productivity levels over time.

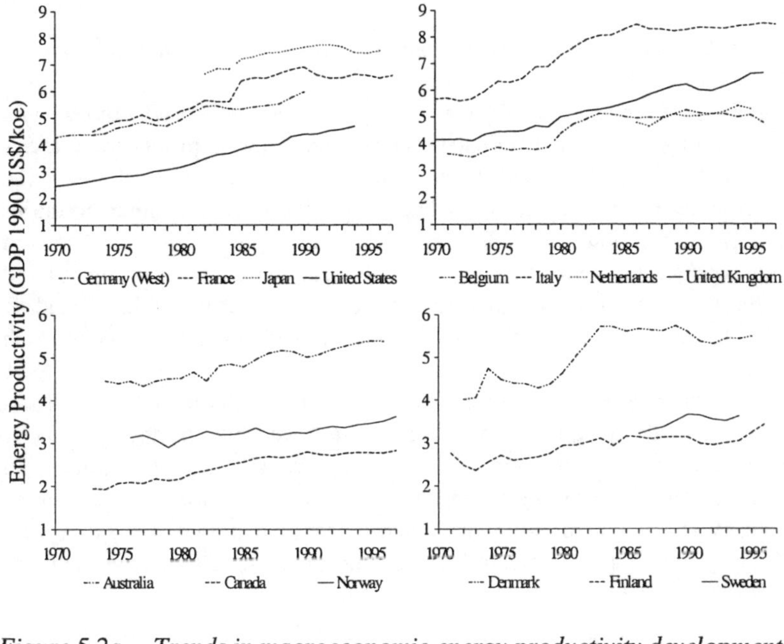

Figure 5.2a Trends in macroeconomic energy productivity development

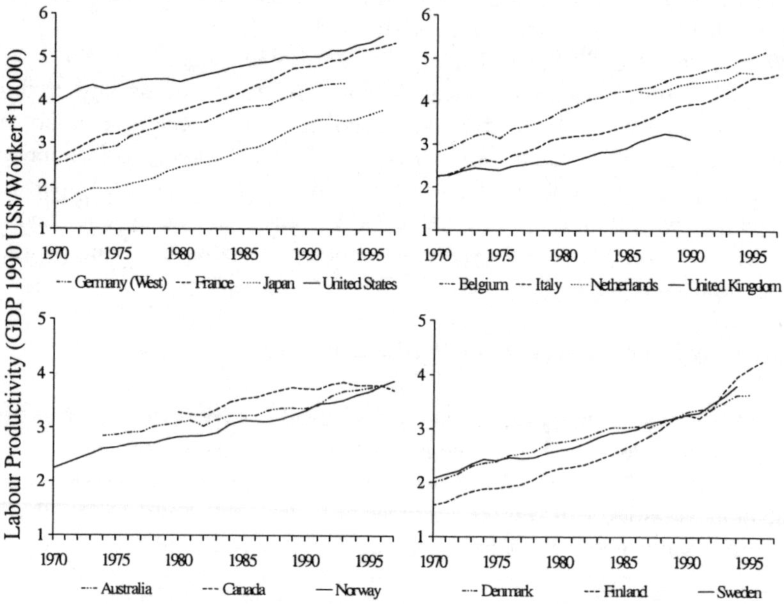

Figure 5.2b Trends in macroeconomic labour productivity development

Figure 5.2a reveals a diverse picture for energy productivity with substantial cross-country differences. The highest energy productivity levels are to be found in Italy and Japan, while Canada, Finland, Norway and Sweden show the lowest levels of energy productivity. All other countries form a medium group. The USA tends to leave the group with low levels of energy productivity over time to catch up with the medium group. Figure 5.2b shows the well-known picture for labour productivity with a leading position for the USA, and other OECD countries showing a tendency to catch up.

These macroeconomic pictures raise the question whether a similar cross-country productivity pattern can be found at a lower level of aggregation, or whether a country's performance differs substantially across sectors. To answer this question we present in Table 5.2 a cross-country comparison of the energy and labour productivity levels relative to the USA for the years 1976 and 1990, for the manufacturing, transport, services and agriculture sectors.

Table 5.2 *Energy and labour productivity relative to USA (USA=100)*

	Manufacturing				Services				Transport				Agriculture			
	Energy		Labour		Energy		Labour		Energy		Labour		Energy		Labour	
	1976	1990	1976	1990	1976	1990	1976	1990	1976	1990	1976	1990	1976	1990	1976	1990
United States	100	100	100	100	100	100	100	100	100	100	100	100	100	100	100	100
Australia	105	71	57	54	413	244	70	72	183	175	57	74	165	91	83	58
Belgium	84	93	59	84	149	130	86	91	481	372	94	124	114	81	79	88
Canada	67	51	74	69	61	52	81	77	115	125	--	74	146	50	89	63
Denmark	188	185	54	49	183	94	66	77	333	316	63	76	34	63	41	64
Finland	57	55	43	64	321	141	50	69	319	268	52	67	145	64	57	61
France	214	172	76	85	74	123	91	106	256	242	65	87	176	135	52	74
Germany (West)	209	175	78	73	106	104	77	98	252	209	55	73	142	103	39	49
Italy	150	166	55	74	1481	598	90	93	370	313	53	67	288	124	41	37
Japan	--	169	--	75	192	151	50	72	662	555	60	84	--	66	30	27
Netherlands	--	75	--	86	--	305	--	89	--	272	--	77	--	38	84	102
Norway	75	42	56	52	102	61	65	67	570	488	78	114	196	114	66	50
Sweden	90	78	52	56	--	63	67	73	182	221	35	52	120	89	59	62
United Kingdom	157	159	53	62	152	152	62	61	368	266	47	58	124	154	66	72
SD log	0.44[a]	0.52[a]	0.22[a]	0.21[a]	0.84[b]	0.61[b]	0.21[c]	0.17[c]	0.55[c]	0.47[c]	0.28[d]	0.25[d]	0.49[a]	0.32[a]	0.35	0.35

Notes:
[a] Excluding Japan and The Netherlands
[b] Excluding The Netherlands and Sweden
[c] Excluding The Netherlands
[d] Excluding Canada and The Netherlands

Concerning energy productivity performance we find that the high energy productivity level of Italy is mainly due to its high energy productivity in services.[6] Japan shows a relatively high level of energy productivity in all four sectors, while Canada, having an overall low level of energy productivity, displays a relatively low level of energy productivity in all four sectors. The overall picture however is that within most countries the energy productivity performance can differ substantially among sectors. For example Finland has a low energy productivity level in manufacturing, while the opposite is true for services. Moreover we find the USA to have an average level of energy productivity in manufacturing while in transport, services and agriculture the USA faces an energy productivity disadvantage relative to most other OECD countries. Concerning labour productivity, Table 5.2 shows that the leading position of the USA holds for all four sectors, and although less pronounced than for energy productivity, there are also substantial cross-sectoral differences within most countries in terms of relative labour productivity performance.

The standard deviation of the log of relative energy and labour productivity performance in Table 5.2 confirms that the cross-country dispersion of energy productivity levels is substantially larger than the cross-country dispersion of relative labour productivity levels.[7] In terms of energy productivity the largest cross-country differences are to be found in services, while agriculture exhibits the largest spread in cross-country labour productivity levels. Finally, cross-country dispersion of both relative energy and labour productivity levels is decreasing over time, with two exceptions: aggregate manufacturing shows a pattern of increasing cross-country differences in energy productivity levels while in agriculture the relative cross-country differences in labour productivity levels remain constant. We explore this issue more extensively in the next chapter.

Decomposing Energy and Labour Productivity Growth Rates

As noted previously, overall productivity performance is not only the result of technology-driven productivity performance in individual sectors, but also of the distribution of production factors among sectors. Therefore we will correct trends in aggregate energy and labour productivity performance for the impact of shifts in sectoral energy and employment shares, to get a better view on the role of sector-specific technology-driven productivity improvements in driving aggregate productivity growth. We do so by using a decomposition- or shift-share analysis, which is based on the following definitions of, respectively, aggregate energy productivity and labour productivity:

$$\frac{Y_t}{E_t} = \sum_i \frac{Y_{i,t}}{E_{i,t}} \frac{E_{i,t}}{E_t} \tag{5.1}$$

$$\frac{Y_t}{L_t} = \sum_i \frac{Y_{i,t}}{L_{i,t}} \frac{L_{i,t}}{L_t} \tag{5.2}$$

with Y_t, E_t and L_t being respectively GDP, final energy consumption and total employment, and the subscript i denoting the subsector. So equation (5.1) says that aggregate energy productivity is the sum of the energy productivity of each subsector (the first term at RHS) multiplied by the energy share of each subsector (the second term at RHS). Equation (5.2) defines the same relationship in terms of labour productivity. Building upon equations (5.1) and (5.2), we decompose aggregate energy and labour productivity growth into a structural effect and an efficiency effect. The structural effect is obtained by calculating aggregate energy and labour productivity growth insofar as it is caused by shifts in sectoral energy and employment shares (the second term at RHS), keeping the levels of energy and labour productivity performance for each individual subsector (the first term at RHS) constant. Vice versa, the efficiency effect is obtained by calculating aggregate energy and labour productivity growth insofar as it is caused by changes in the energy and labour productivity performance within each individual subsector, keeping the sectoral energy and employment shares constant. Hence the structural effect indicates the effect of changes in the structure of production on aggregate productivity growth, while the efficiency effect points to the role of technology-driven efficiency improvements.

In this chapter we perform a decomposition analysis at two levels of aggregation. In this section we decompose for each of the 14 OECD countries the average annual macroeconomic energy and labour productivity growth rate into a structural and an efficiency effect, examining the role of the manufacturing, services, transport and agriculture sectors. In section 5.4 we repeat the analysis to decompose average annual manufacturing energy and labour productivity growth rates, examining the role of ten manufacturing subsectors in driving energy and labour productivity trends in aggregate manufacturing.

Many studies measure the relative contribution of structural and technological change to aggregate productivity growth, using so-called index number decomposition or shift-share analysis.[8] The studies differ from each other in several dimensions, including the number of sectors and countries included, the methodology (Laspyeres, Paasche, Divisia, and so on), the area of application (TFP, capital, labour, energy), the type of indicator (quantity, intensity, productivity or elasticity) and the type of analysis (time-series or period-wise). For a lucid exposition of the methodology and a survey of studies we refer to Ang (1995a, 1995b, 1999) and Ang and Zhang (2000) concerning energy studies, and to Syrquin (1984) concerning macroeconomic studies focusing on aggregate (total factor) productivity. The main value added of our study lies in a simultaneous exploration of productivity performance along the two dimensions of energy and labour. Moreover compared to most existing studies our analysis comprises a relatively high level of sectoral detail for a relatively large number of countries, in particular in terms of energy productivity developments. As a result, the changes in technology-driven productivity performance at the level of individual sectors reported in this chapter are relatively well specified and informative. Furthermore, contrary to most studies, in particular those focusing on energy productivity, our decomposition analysis is not confined to the manufacturing sector, but applies also to the macroeconomic level, identifying the role of manufacturing, services, transport and agriculture in driving macroeconomic productivity growth rates. Finally, our data set enables us to apply a time-series approach whereas most cross-country studies conduct a period-wise approach, using only data for the first and the last year of a specified time period. Compared to a period-wise approach, a time-series approach yields more insight into energy productivity development over subsequent years, and moreover the decomposition results are less sensitive to the exact functional form used and to the values in the initial and final year.

Several functional forms can be used for the actual decomposition. We use the so-called Refined Divisia Index method and refer to Appendix C for a motivation as well as technical details and a brief discussion of alternative decomposition methods.

In Figures 5.3a and 5.3b we present the results of the decomposition of the macroeconomic energy and labour productivity growth rates into a structural effect and an efficiency effect. Figures 5.3a and 5.3b plot for each country, respectively, the average annual macroeconomic energy and labour productivity growth rate as the sum of an efficiency effect and a structural effect. It is to be noted that one has to be careful with comparing the results between countries due to the different time periods used (because of data availability). From Figure 5.3a it can be seen that except for Belgium, Sweden, the United Kingdom and the United States, structural changes explain a substantial part of average annual macroeconomic energy productivity growth rates. Structural effects even dominate efficiency effects in Australia, Denmark, Finland, Italy, the Netherlands and Norway. In most countries the efficiency effect is positive, except for Finland, Italy and the Netherlands.

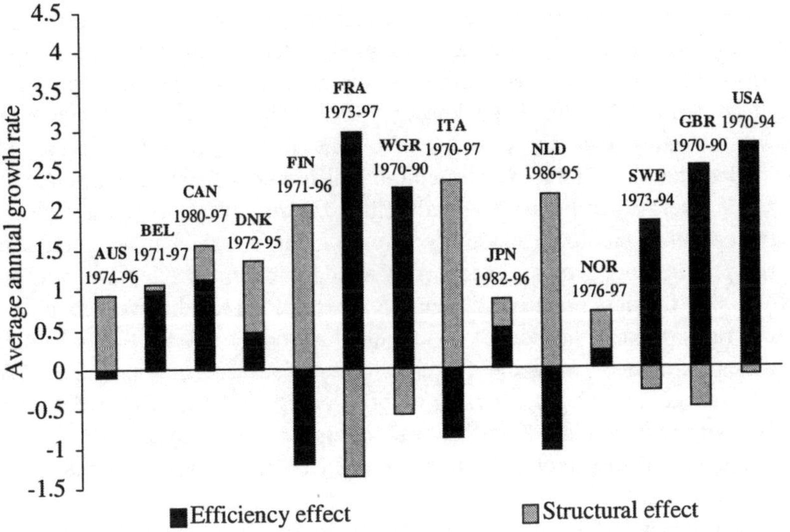

Figure 5.3a Decomposition of average annual growth rate of macroeconomic energy productivity

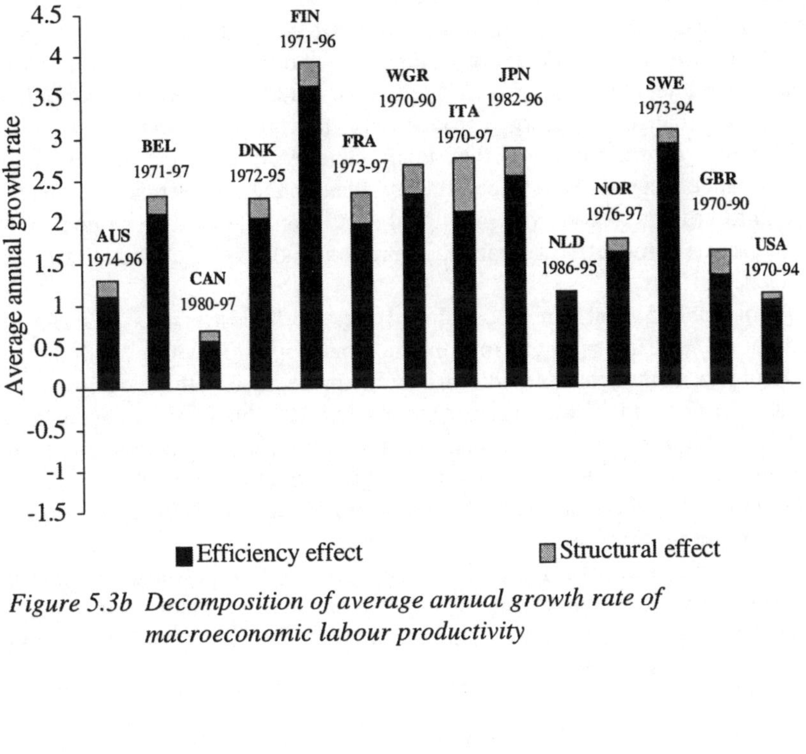

Figure 5.3b Decomposition of average annual growth rate of macroeconomic labour productivity

Figure 5.3b shows on the contrary that although in all countries the effect of structural changes on macroeconomic labour productivity growth rates is positive, it is also relatively small, implying efficiency improvements to be the main source of macroeconomic labour productivity growth. The latter result confirms what has been known from the macroeconomic empirical growth literature (see for example van Ark 1996). Moreover, it can be concluded that considerable cross-country differences exist, in particular in terms of energy productivity. Finally, the figure reveals that on average macroeconomic labour productivity growth is higher than macroeconomic energy productivity growth, except for Canada, the United Kingdom and the USA. Using the data underlying Figure 5.3, we calculated the average annual growth rates of energy productivity and labour productivity for the 14 OECD countries combined, weighted for each country's share in total GDP. We found average annual growth rates of both energy productivity and labour productivity to be about 1.8 per cent before correcting for structural changes, while they are, respectively, 1.7 per cent and 1.6 per cent after correcting for structural changes.

To see which sectors are responsible for these aggregate results, in Table 5.3 we split the percentage contribution of the total efficiency effect and the total structural effect to the aggregate productivity growth rates, as presented in Figure 5.3, into the percentage contribution of individual subsectors. In Table 5.3a, for each country the first column denotes per individual sector its shift in energy share, expressed as a percentage contribution to the total effect of shifts in sectoral energy shares on aggregate productivity growth (that is, the total structural effect). The second column denotes per individual sector its change in energy productivity performance, expressed as a percentage contribution to the total change in energy productivity performance at a constant sector structure (that is, the total efficiency effect). The third column denotes per individual sector its total relative contribution to aggregate productivity change, being the sum of the structural and efficiency effects.

From Table 5.3a it can be concluded that the largest effects of shifts in sectoral energy shares on macroeconomic energy productivity growth are to be found in manufacturing and services, with the energy share declining in manufacturing and increasing in services (except for West Germany and Sweden). Moreover it can be seen that the extraordinary positive effect of structural changes on macroeconomic energy productivity growth in Finland, Italy and the Netherlands is to be explained by an exceptional increasing energy share in services.[9]

Finally, the effect of shifts in the energy share of transport and agriculture on macroeconomic structural change is relatively small, with small increasing energy shares in transport and a mix of increasing and decreasing energy shares in agriculture (decreasing in Denmark, Finland, France, West Germany, Sweden, the United Kingdom and the USA and increasing in other countries). Concerning energy efficiency improvements, Table 5.3a shows that they are mainly realized within manufacturing. For services however the picture is highly diverse with a mix of positive and negative percentage contributions to aggregate energy efficiency improvements. Most notable is again the exceptional negative growth rate of energy productivity in Finland, Italy and the Netherlands, which drive the negative efficiency effects in these countries as plotted in Figure 5.3a. The percentage contribution of transport and agriculture to macroeconomic energy efficiency improvements is relatively small (except for Norway), with energy efficiency improving in transport (except for Belgium, West Germany, Japan, Norway and the United Kingdom) while energy efficiency in agriculture slightly improves in Australia, Denmark, France, West Germany, Sweden, the United Kingdom and the USA, and slightly decreases in the other countries.

In Table 5.3b we present a similar breakdown of the total structural and efficiency effects as in Table 5.3a, but for labour productivity. Table 5.3b shows that the relatively small impact of total structural change on macroeconomic labour productivity growth does not imply that employment mixes have been constant over time. On the contrary the employment mix changed considerably with a substantial decreasing employment share in manufacturing and a substantially increasing employment share in services. The fact that the net effect of this shift on macroeconomic labour productivity growth is always positive confirms an employment shift from a relatively low- towards a relatively high value added sector.

Moreover Table 5.3b also shows that in terms of shifts in employment shares, the relative contribution of transport and agriculture to macroeconomic structural change is small, with decreasing employment shares in agriculture and a mix of increasing and decreasing employment shares in transport (decreasing in Australia, Canada, the Netherlands, Norway, the United Kingdom and the USA, constant in Belgium, and increasing in other countries). Concerning the efficiency effect, manufacturing is not only an important source for energy efficiency improvement, but also for labour efficiency improvement (that is, labour productivity corrected for structural changes). Moreover, unlike energy efficiency, services is also an important source for labour efficiency improvement in most countries, except for the Netherlands. Similar to energy efficiency, the percentage contributions of transport and agriculture to macroeconomic labour efficiency improvements are small, although positive in all countries.

Table 5.3a Percentage contribution of the efficiency effect (EFF) and the structural effect (STR) by sector to average annual growth rate (g) of aggregate energy productivity per country

	Australia 1974–96			Belgium 1971–97			Canada 1980–97			Denmark 1972–95			Finland 1971–95		
	STR	EFF	Total	STR	EFF	Total	STR	EFF	Total	STR	EFF	Total	STR	EFF	Total
Manufacturing	-20.9	5.6	-15.3	-27.6	53.6	26.0	-1.1	18.9	17.9	-47.4	65.5	18.1	-19.6	89.9	70.3
Transport	4.1	8.4	12.4	13.1	-9.7	3.4	-1.2	4.6	3.4	4.9	10.0	14.9	5.9	1.8	7.7
Agriculture	0.7	0.2	0.9	9.0	-8.9	0.1	5.3	-5.0	0.4	-3.6	11.7	8.1	-9.9	-27.1	-37.0
Services	126.3	-24.3	101.9	12.6	57.9	70.5	24.1	54.3	78.4	112.0	-53.0	58.9	259.0	-200.0	59.0
Total per cent	110.2	-10.2	**100.0**	7.1	92.9	**100.0**	27.1	72.9	**100.0**	65.9	34.1	**100.0**	235.5	-135.5	**100.0**
Total *g*	0.94	-0.09	**0.86**	0.08	1.01	**1.08**	0.42	1.14	**1.56**	0.90	0.47	**1.37**	1.53	-0.88	**0.65**

	France 1985–97			Germany (West) 1970–90			Italy 1990–97			Japan 1982–96			The Netherlands 1986–95		
	STR	EFF	Total	STR	EFF	Total	STR	EFF	Total	STR	EFF	Total	STR	EFF	Total
Manufacturing	-103.8	61.4	-42.3	-24.3	38.7	14.4	-30.1	62.4	32.3	-38.7	67.0	28.3	-41.3	46.9	5.6
Transport	20.2	6.9	27.1	4.9	-0.6	4.2	6.5	0.5	7.0	9.8	-1.0	8.8	8.7	6.8	15.5
Agriculture	-27.9	10.9	-17.0	-0.9	1.9	1.0	3.9	-7.4	-3.5	3.3	-14.0	-10.8	14.4	-4.5	10.0
Services	2.0	130.2	132.2	-13.7	94.1	80.3	178.9	-114.7	64.1	68.3	5.3	73.6	210.3	-141.4	69.0
Total per cent	-109.4	209.4	**100.0**	-34.1	134.1	**100.0**	159.2	-59.2	**100.0**	42.7	57.3	**100.0**	192.1	-92.1	**100.0**
Total *g*	-0.26	0.50	**0.24**	-0.58	2.27	**1.70**	2.36	-0.88	**1.48**	0.37	0.50	**0.87**	2.18	-1.05	**1.14**

Table 5.3a Continued

	Norway 1976–97			Sweden 1973–94			United Kingdom 1970–90			United States 1970–94		
	STR	EFF	Total	STR	EFF	Total	STR	EFF	Total	STR	EFF	Total
Manufacturing	-29.7	-4.7	-34.3	-1.9	25.4	23.5	-34.7	47.1	12.3	-12.1	29.6	17.5
Transport	22.0	38.4	60.4	2.6	9.7	12.3	9.6	-2.4	7.3	1.6	2.5	4.1
Agriculture	28.9	-29.8	-0.9	-11.3	10.6	-0.8	-2.0	4.4	2.4	-0.6	3.0	2.4
Services	47.1	27.7	74.8	-8.6	73.6	65.0	2.6	75.4	78.0	7.5	68.5	76.0
Total per cent	68.4	31.6	**100.0**	-19.2	119.2	**100.0**	-24.5	124.5	**100.0**	-3.5	103.5	**100.0**
Total *g*	0.48	0.22	**0.70**	-0.30	1.84	**1.54**	-0.50	2.52	**2.03**	-0.10	2.81	**2.72**

Table 5.3b *Percentage contribution of the efficiency effect (EFF) and the structural effect (STR) by sector to average annual growth rate (g) of aggregate labour productivity per country*

	Australia 1974–96			Belgium 1971–97			Canada 1980–97			Denmark 1972–95			Finland 1971–95		
	STR	EFF	Total	STR	EFF	Total	STR	EFF	Total	STR	EFF	Total	STR	EFF	Total
Manufacturing	-32.4	28.8	-3.6	-23.7	50.0	26.3	-50.9	4.9	-46.0	-3.7	25.6	21.9	-1.5	42.9	41.4
Transport	-3.0	13.2	10.2	0.0	6.2	6.2	-11.1	15.3	4.2	2.2	11.2	13.4	2.4	6.0	8.4
Agriculture	-4.6	6.6	2.0	-3.1	4.6	1.5	-11.6	10.1	-1.5	-7.6	14.8	7.3	-8.3	11.8	3.5
Services	54.6	36.8	91.4	36.2	29.7	66.0	91.1	52.1	143.2	19.6	37.8	57.4	14.6	32.1	46.7
Total per cent	14.6	85.4	**100.0**	9.5	90.5	**100.0**	17.5	82.5	**100.0**	10.6	89.4	**100.0**	7.2	92.8	**100.0**
Total *g*	0.19	1.12	**1.31**	0.22	2.10	**2.32**	0.12	0.57	**0.70**	0.24	2.04	**2.29**	0.28	3.65	**3.94**

	France 1985–97			Germany (West) 1970–90			Italy 1990–97			Japan 1982–96			The Netherlands 1986–95		
	STR	EFF	Total	STR	EFF	Total	STR	EFF	Total	STR	EFF	Total	STR	EFF	Total
Manufacturing	-24.2	45.0	20.8	-12.6	38.0	25.3	-11.6	42.3	30.8	-4.9	36.3	31.4	-49.6	55.0	5.3
Transport	2.1	5.7	7.8	0.6	3.8	4.3	1.5	5.0	6.5	0.8	7.5	8.3	-1.1	16.7	15.6
Agriculture	-9.3	11.5	2.2	-3.7	5.1	1.4	-6.6	7.4	0.8	-4.4	3.3	-1.1	-11.2	21.2	10.0
Services	44.4	24.7	69.2	28.8	40.1	68.9	40.1	21.8	61.9	20.1	41.2	61.3	60.0	9.1	69.1
Total per cent	13.0	87.0	**100.0**	13.0	87.0	**100.0**	23.4	76.6	**100.0**	11.7	88.3	**100.0**	-1.9	101.9	**100.0**
Total *g*	0.26	1.74	**2.00**	0.35	2.32	**2.67**	0.64	2.10	**2.75**	0.33	2.53	**2.86**	-0.02	1.14	**1.12**

Table 5.3b Continued

	Norway 1976–97			Sweden 1973–94			United Kingdom 1970–90			United States 1970–94		
	STR	EFF	Total	STR	EFF	Total	STR	EFF	Total	STR	EFF	Total
Manufacturing	-19.0	19.2	0.1	-21.1	49.7	28.6	-57.4	64.2	6.7	-45.3	52.2	6.9
Transport	-6.9	41.5	34.5	2.5	6.6	9.1	-4.5	11.8	7.3	-1.8	5.3	3.5
Agriculture	-8.0	11.3	3.3	-2.6	4.2	1.6	-3.6	6.0	2.4	-4.0	6.0	2.0
Services	42.9	19.1	62.0	26.8	33.8	60.7	83.7	-0.1	83.6	59.1	28.6	87.6
Total per cent	8.9	91.1	**100.0**	5.7	94.3	**100.0**	18.1	81.9	**100.0**	8.0	92.0	**100.0**
Total g	0.16	1.60	**1.76**	0.17	2.89	**3.06**	0.29	1.32	**1.61**	0.09	1.02	**1.10**

Sectoral Biases in Productivity Growth Rates

So far, we have found macroeconomic growth rates of labour productivity in general to be substantially higher than macroeconomic growth rates of energy productivity. Does this pattern hold also for manufacturing, services, transport and agriculture? In this section we take a closer look at the relationship between sectoral growth rates of energy and labour productivity. Are they positively or negatively correlated to one another among the different countries? In other words, do they complement each other, or are they substitutes? And is the observed relationship between energy and labour productivity growth changing over time?

To assess these issues, we calculate the average annual growth rates of energy and labour productivity for each sector and country for the period 1970–97.[10] They are presented in Figure 5.4 together with two regression lines through the origin, estimating the cross-sectional relationship between energy and labour productivity growth rates for the periods 1970–82 and 1982–97.

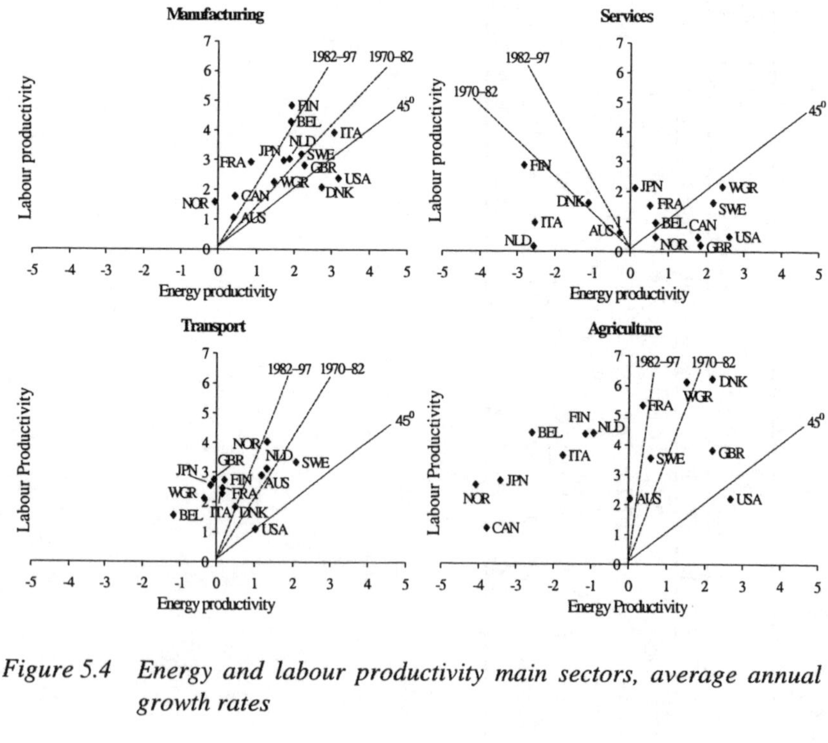

Figure 5.4 Energy and labour productivity main sectors, average annual growth rates

The figure leads to the following conclusions. In manufacturing all countries show a positive correlation between energy and labour productivity growth rates, suggesting manufacturing energy and labour productivity growth to be complements rather than substitutes. For most countries, this conclusion holds also for services and transport. In agriculture, however, seven of the 14 countries combine a positive labour productivity growth with a negative energy productivity growth, suggesting energy and labour productivity growth to be substitutes rather than complements in these countries. Of course the figure shows again that labour productivity growth is in general substantially higher than energy productivity growth.

Comparing the regression lines for the period 1982–97 and the period 1970–82 suggests that this bias towards labour productivity growth is increasing in aggregate manufacturing, transport and agriculture, while it is decreasing in services. Insofar as the observed sectoral productivity growth rates are driven by technological progress, the increasing bias towards labour productivity growth suggests the existence of an increasing bias towards labour-augmenting technological progress in aggregate manufacturing, transport and agriculture.

In the remainder of this chapter we move beyond the macroeconomic level by taking a closer look at the manufacturing sector. As already argued several times, a prerequisite for understanding technology-driven productivity performance is to assess productivity performance at a sufficiently disaggregated sector level. Therefore, we continue by identifying cross-country productivity developments within ten manufacturing subsectors, following the same research strategy we have applied so far.

5.4 THE MANUFACTURING SECTOR IN DETAIL

In this section we further explore levels and trends in manufacturing energy and labour productivity performance, examining the role of ten manufacturing subsectors: food, textiles, wood, paper, chemicals, non-metallic minerals, iron and steel, non-ferrous metals, machinery and transport equipment (see Table 5.1). We start with a brief overview of their sectoral shares in manufacturing energy consumption, employment and GDP (see Figure 5.5). For 12 OECD countries included in this study taken together (excluding Australia and Canada), the subsector chemicals consumed by far the most energy with a share of 40 per cent in manufacturing final energy consumption in 1990, followed by iron and steel (16 per cent), paper, pulp and printing (11 per cent) and non-metallic minerals (9 per cent).[11] In Figure 5.5 we compare those shares with the shares of total employment and value added.[12] This yields a different picture, with machinery accounting for the largest share of manufacturing total employment and value added (35–37 per cent) followed by food and transport equipment (each around 12 per cent).

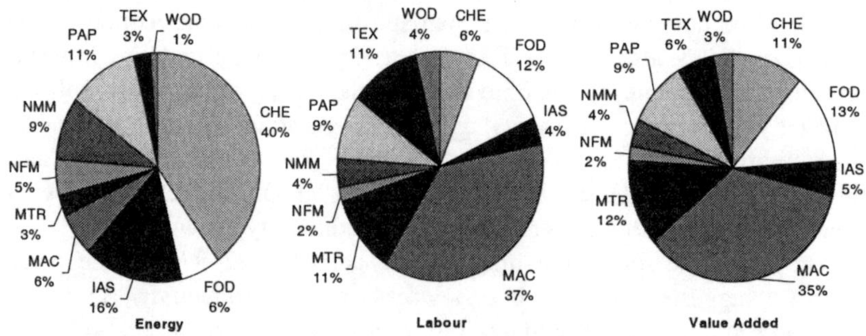

Figure 5.5 Percentage shares of manufacturing final energy consumption, labour and value added. Sample of 12 OECD countries in 1990

In the previous section we found the shares of total energy consumption, employment and value added for aggregate manufacturing to be more or less similar among the 14 OECD countries (see Table 5.A.1 in Appendix 5.A). Within manufacturing however these shares differ substantially among the different manufacturing subsectors. For example, in the Netherlands, the energy-intensive chemicals sector is responsible for 67 per cent of manufacturing energy consumption, while in the other countries this share lies between 14 per cent and 38 per cent. In Finland and Sweden, paper consumes around 40 per cent of manufacturing energy consumption, while in the other countries this share lies in between 3 per cent and 14 per cent. Contrary to other countries, in Italy textiles is responsible for a large share of total employment and value added. For a detailed overview of manufacturing sector shares per country we refer to Table 5.A.3 in Appendix 5.A.

Comparing Energy and Labour Productivity Levels

To compare cross-country energy and labour productivity performance at the manufacturing level, we show in Figure 5.6 the aggregate manufacturing energy and labour productivity levels over time, for each of the 14 OECD countries. Figure 5.6a shows that in aggregate manufacturing two groups of countries can be identified in terms of observed levels of energy productivity. Denmark, Italy, West Germany, France, Japan and the United Kingdom, show a high level of energy productivity while the USA, Netherlands, Belgium, Sweden, Australia, Finland, Canada and Norway display a relatively low energy productivity level.[13]

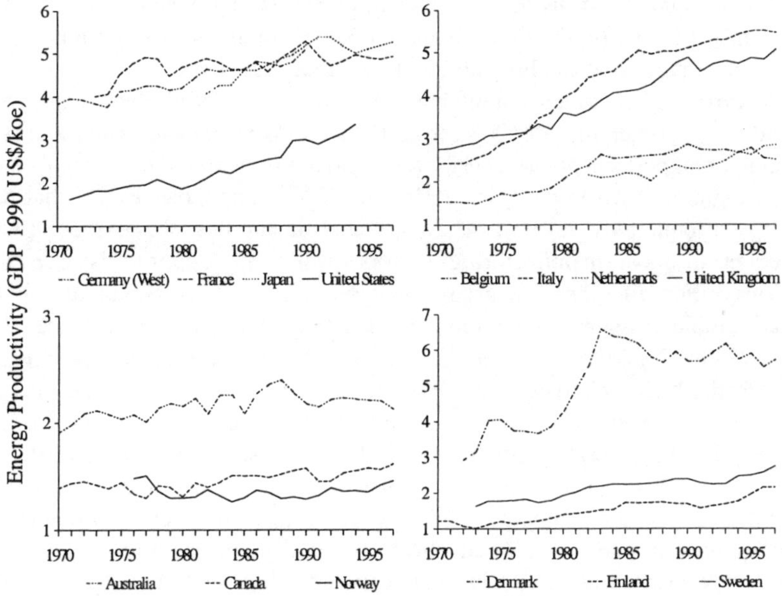

Figure 5.6a Trends in manufacturing energy productivity development

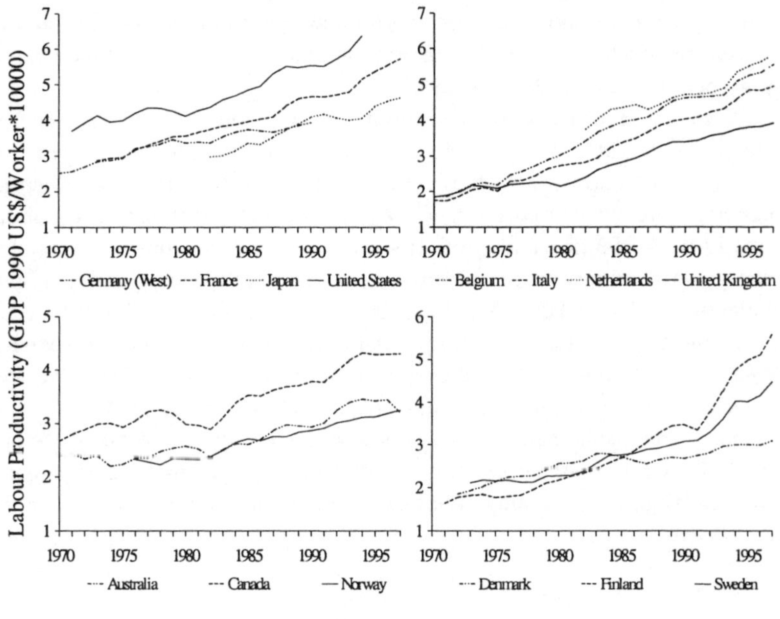

Figure 5.6b Trends in manufacturing labour productivity development

Figure 5.6b, confirms again the well-known leading position of the USA in terms of labour productivity, with – contrary to the macroeconomic level – no clear pattern of catching up by other OECD countries.

In order to see which manufacturing sectors are driving these aggregate trends, we provide in Table 5.4 for all ten manufacturing subsectors a cross-country comparison of the energy and labour productivity level relative to the weighted OECD average in 1976 and 1990.[14] The table reveals that the energy productivity level in Germany and Japan lies above the OECD average in most manufacturing sectors, while the opposite is true for Norway. For all other countries, the table shows a diverse picture with considerable cross-sector variation in relative productivity performance. For example the Netherlands has a relatively low level of energy productivity in chemicals, but a relatively high level in paper and wood. The high energy productivity level in Denmark, as shown in Figure 5.6a, is due to an extremely high energy productivity level in chemicals and paper, while its energy productivity level in food is relatively low. A few other remarkable facts are: Finland, Norway and Sweden have low levels of energy productivity in paper and wood; the United Kingdom has a relatively high level of energy productivity in chemicals, iron and steel, and paper, while this is relatively low in machinery; and Italy has a very high level of energy productivity in wood.

Concerning labour productivity, Table 5.4 again confirms the well-known leading position of the USA for most manufacturing sectors. Exceptions however are non-ferrous metals and non-metallic minerals, where the USA is lagging behind some other countries.

Looking at the standard deviation of the log of relative energy productivity in Table 5.4 leads to the conclusion that also at the level of manufacturing subsectors the cross-country differences in energy productivity are substantially larger than cross-country differences in labour productivity. In the previous section we found the cross-country dispersion of energy productivity to be increasing over time at the level of aggregate manufacturing. From Table 5.4 it can be concluded however that this result does not apply to all manufacturing sectors: we find cross-country dispersion of energy productivity to be increasing only in the energy-intensive sectors of chemicals, iron and steel and non-ferrous metals, while it is more or less constant in chemicals, paper and wood, and decreasing in the other sectors. Overall, the cross-country differences in labour productivity performance seem to be slightly decreasing. We turn to this issue more extensively in the next chapter.

Table 5.4 Energy and labour productivity relative to average OECD (OECD=100)

	Chemicals				Food				Iron and steel				Machinery				Transport equipment			
	Energy		Labour		Energy		Labour		Energy		Labour		Energy		Labour		Energy		Labour	
	1976	1990	1976	1990	1976	1990	1976	1990	1976	1990	1976	1990	1976	1990	1976	1990	1976	1990	1976	1990
OECD*	100	100	100	100	100	100	100	100	100	100	100	100	100	100	100	100	100	100	100	100
Belgium	97	149	70	78	116	115	85	97	76	58	46	80	98	118	98	94	153	134	81	106
Canada	96	64	96	92	171	--	103	87	121	55	96	76	201	--	99	86	92	--	92	93
Denmark	269	434	70	53	40	67	45	61	74	96	31	54	89	116	84	55	76	91	65	54
Finland	68	88	47	55	49	74	51	65	83	50	39	69	59	94	58	82	131	161	--	63
France	149	168	63	77	125	120	95	89	169	129	57	63	205	100	102	98	198	193	81	90
Germany (West)	202	175	84	63	135	162	92	86	163	127	67	65	299	215	111	85	91	111	99	89
Italy	54	141	32	62	87	168	84	107	165	128	72	96	72	102	83	95	319	335	63	91
Japan	155	176	132	113	240	226	92	76	155	154	139	137	--	179	40	91	252	171	66	102
Netherlands	48	66	79	89	59	76	70	89	120	70	97	85	--	100	--	90	195	106	65	74
Norway	55	58	47	64	57	54	69	48	38	18	43	55	69	61	81	66	83	69	60	58
Sweden	101	147	73	53	78	90	73	67	77	85	31	48	57	68	65	64	86	96	--	57
United Kingdom	166	230	66	73	84	114	68	81	169	130	42	62	64	68	71	65	79	142	51	83
United States	--	--	113	124	--	--	125	129	--	--	115	100	--	--	123	125	--	--	135	113
SD log	0.54[a]	0.56[a]	0.36	0.27	0.50[b]	0.42[b]	0.27	0.25	0.45[a]	0.58[a]	0.48	0.28	0.55[c]	0.36[c]	0.30[d]	0.22[d]	0.49[b]	0.41[b]	0.26[e]	0.22[e]

Notes:
* Average OECD is weighted with each country's 1990 GDP share of total GDP per sector.
[a] Excluding the United States
[b] Excluding Canada and the United States
[c] Excluding Canada, Japan, the Netherlands and the United States
[d] Excluding the Netherlands
[e] Excluding Finland and Sweden

Table 5.4 Continued

	Non-ferrous metals				Non-metallic minerals				Paper				Textiles				Wood			
	Energy		Labour		Energy		Labour		Energy		Labour		Energy		Labour		Energy		Labour	
	1976	1990	1976	1990	1976	1990	1976	1990	1976	1990	1976	1990	1976	1990	1976	1990	1976	1990	1976	1990
OECD*	100	100	100	100	100	100	100	100	100	100	100	100	100	100	100	100	100	100	100	100
Belgium	143	156	126	165	46	97	79	119	265	227	55	84	328	124	82	108	145	145	39	44
Canada	87	31	171	113	157	138	136	103	54	30	88	85	--	--	107	91	130	--	100	86
Denmark	--	--	46	48	93	76	92	70	611	322	74	61	130	112	95	74	129	102	95	56
Finland	111	77	80	103	68	63	65	86	33	33	47	84	176	145	66	69	41	56	65	77
France	133	125	174	159	222	143	126	137	341	182	82	93	151	170	118	116	--	--	83	91
Germany (West)	196	163	117	78	148	139	107	100	290	188	68	76	149	108	107	98	185	217	111	75
Italy	113	108	99	102	119	136	93	98	229	262	59	103	191	216	115	109	803	756	89	80
Japan	249	193	227	152	121	135	63	88	--	185	66	98	85	115	55	46	--	--	48	52
Netherlands	111	51	256	168	115	118	--	111	360	288	72	93	131	134	108	108	402	300	134	89
Norway	35	20	137	96	65	88	100	79	83	56	56	62	90	95	62	66	87	62	90	66
Sweden	112	72	92	72	86	96	84	82	37	31	58	71	112	64	105	82	76	67	109	87
United Kingdom	100	98	67	66	123	128	79	75	290	374	65	82	106	139	91	83	--	--	88	59
United States	--	--	140	113	--	--	118	105	--	--	126	112	--	--	99	110	--	--	124	106
SD log	0.47[f]	0.68[f]	0.47	0.37	0.41[a]	0.26[a]	0.23	0.19	0.97[g]	0.97[g]	0.24	0.18	0.36[b]	0.30[b]	0.24	0.26	0.89[h]	0.85[h]	0.34	0.25

Notes:
* Average OECD is weighted with each country's 1990 GDP share of total GDP per sector.
[f] Excluding Denmark and the United States
[g] Excluding Japan and the United States
[h] Excluding Canada, France, Japan, the United Kingdom and the United States

Decomposing Energy and Labour Productivity Growth Rates

To get a more precise view of the role of sector-specific technology-driven productivity improvements in driving the observed trends in manufacturing energy and labour productivity performance, we have to correct the latter for the impact of shifts in sectoral energy and employment shares. Hence in this section we decompose average annual manufacturing energy and labour productivity growth rates into a structural and an efficiency effect, examining the role of the ten manufacturing subsectors. Again we use the Refined Divisa Method for the actual decomposition (see Appendix C). The results are presented in Figures 5.7a and 5.7b.[15] Figure 5.7a shows that in all 12 OECD countries energy efficiency improvements are the main driving force behind aggregate manufacturing energy productivity growth, although in most countries there is also a substantial effect from shifts in sectoral energy shares on aggregate manufacturing energy productivity growth. This structural effect is mixed: it is positive in Belgium (41 per cent), Denmark (11 per cent), France (47 per cent), West Germany (20 per cent), Italy (37 per cent), Japan (33 per cent) and the USA (35 per cent), indicating a shift towards a less energy-intensive manufacturing structure, while it is negative in Finland (-50 per cent), the Netherlands (-30 per cent), Norway (-960 per cent) and Sweden (-12 per cent), indicating a shift towards a more energy-intensive manufacturing structure. In Norway the large structural change even dominates the energy efficiency improvements.

This overall picture accords well with other cross-country studies decomposing manufacturing energy use in OECD countries (Greening et al. 1997, Howarth et al. 1991, Eichhammer and Mannsbart 1997 and Unander et al. 1999), although our structural effects in Finland, France and Italy are relatively high as compared to these studies. This might well be due to differences in data, period and decomposition method between the other studies and ours.[16] Concerning labour productivity growth, Figure 5.7b shows that in all 12 OECD countries the effect of shifts in sectoral employment shares on aggregate manufacturing labour productivity growth is positive, but also very small; almost all aggregate manufacturing labour productivity growth is to be explained from labour productivity improvements in individual sectors. This result confirms what has been known from empirical labour productivity analyses for the manufacturing sector (see, for example, Dollar and Wolff 1993, Fagerberg 2000).

Similar to the conclusions drawn at the macroeconomic level, considerable cross-country differences also exist at the level of manufacturing, in particular in terms of energy productivity.

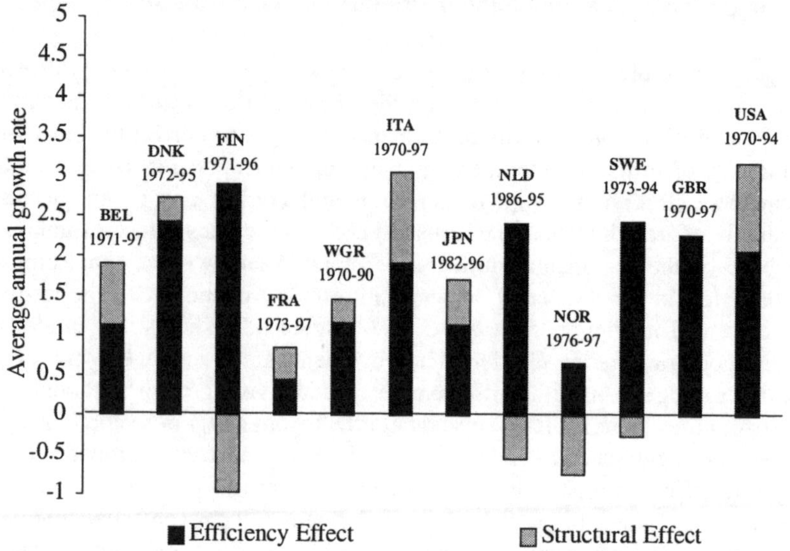

Figure 5.7a Decomposition of average annual growth rate of manufacturing energy productivity

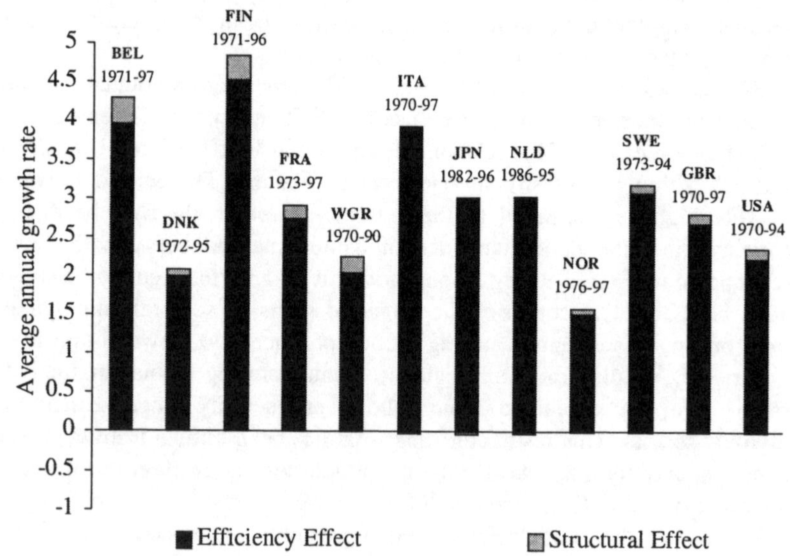

Figure 5.7b Decomposition of average annual growth rate of manufacturing labour productivity

Moreover except for Denmark and the USA, the average macroeconomic labour productivity growth is again higher than macroeconomic energy productivity growth. We calculated that for the 12 OECD countries taken together, the weighted average annual growth rates of manufacturing energy and labour productivity are, respectively, 2.25 per cent and 2.69 per cent, while they drop to, respectively, 1.57 per cent and 2.53 per cent after being corrected for the impact of structural changes.

To see which sectors are responsible for these aggregate results, in Table 5.5a we split the percentage contribution of the total efficiency and structural effects to aggregate manufacturing energy productivity growth rates, as presented in Figure 5.7a, into the percentage contribution of individual subsectors for each country. The interpretation of the figures is similar to Table 5.3a.

From Table 5.5a it can be concluded that except for France, the aggregate manufacturing energy productivity improvements are to a large extent realised within the machinery sector, followed by chemicals.[17] Looking into the sources of structural changes yields a more diverse picture.

In Belgium, the substantial positive structural effect on aggregate manufacturing energy productivity growth is mainly caused by a shift of energy share from machinery, iron and steel, non-ferrous metals and non-metallic minerals towards chemicals, food, transport equipment and paper.

The small positive structural effect in Denmark is mainly the result of a relatively small increasing energy share in food and machinery and decreasing energy shares in textiles, transport equipment, non-metallic minerals and iron and steel, while the role of non-ferrous metals is unclear due to lack of data.

The substantial negative impact of structural changes on aggregate manufacturing energy productivity growth in Finland is to a large extent caused by a shift in energy share from machinery, food, transport equipment and textiles towards paper, non-specified industry, iron and steel and chemicals.

In France the positive structural effect is mainly due to a shift of energy shares from textiles, iron and steel and transport equipment towards machinery, paper and food.

In West Germany the positive structural effect is mainly caused by an increasing energy share in machinery and chemicals at the cost of a decreasing energy share in iron and steel, textiles and non-metallic minerals.

The positive structural effect in Italy has been mainly due to a shift of energy shares from non-metallic minerals and iron and steel towards machinery, non-specified industry, food and textiles.

Table 5.5a Percentage contribution of the structural effect (STR) and the efficiency effect (EFF) by sector to average annual growth rate (g) of aggregated energy productivity in manufacturing per country

	Belgium 1971–97			Denmark 1972–97			Finland 1971–97			France 1973–97			Germany (West) 1970–90			Italy 1970–97		
	STR	EFF	Total	STR	EFF	Total	STR	EFF	Total	STR	EFF	Total	STR	EFF	Total	STR	EFF	Total
Chemicals	13.5	14.9	28.5	-2.7	20.7	18.0	1.4	5.2	6.6	8.4	25.3	33.7	12.6	3.4	16.0	-0.8	10.8	10.0
Food	15.6	-3.8	11.8	10.9	10.6	21.4	-13.6	13.5	-0.1	17.8	2.2	20.0	0.1	5.0	5.1	4.7	5.5	10.2
Iron and steel	-4.8	4.7	-0.1	-1.3	2.8	1.5	2.1	4.6	6.6	-12.5	9.6	-2.9	-9.1	9.3	0.2	-1.0	1.6	0.6
Machinery	-24.4	39.2	14.7	7.3	31.1	38.5	-17.0	88.2	71.2	66.7	-36.4	30.3	16.3	20.0	36.3	9.8	22.5	32.3
Transport equipment	16.5	-5.4	11.1	-4.8	1.5	-3.3	-11.6	12.0	0.4	-10.9	13.9	3.0	9.6	15.7	25.2	6.4	-3.5	2.9
Non-ferrous metals	-1.7	5.5	3.8	--	--	--	0.6	1.3	1.9	-5.1	7.9	2.8	1.5	2.7	4.2	0.1	0.4	0.5
Non-metallic minerals	-3.6	5.9	2.3	-3.1	1.5	-1.7	-5.3	5.8	0.5	0.3	-2.9	-2.5	-3.2	5.2	2.1	-2.1	8.8	6.7
Paper	12.6	-6.6	6.0	1.0	5.7	6.7	5.3	9.9	15.2	20.1	-13.1	7.0	4.2	0.0	4.2	3.8	3.8	7.6
Textiles	10.2	-12.4	-2.2	-7.1	6.7	-0.4	-12.4	4.7	-7.7	-37.4	17.5	-19.9	-4.3	-0.8	-5.0	4.7	9.7	14.4
Wood	1.0	0.2	1.2	3.1	-0.3	2.9	-6.2	7.2	1.0	--	--	--	-1.0	1.6	0.5	1.8	-1.4	0.4
Non-specified industry	5.9	17.0	22.9	7.8	8.6	16.4	5.9	-1.4	4.4	0.1	28.4	28.5	-6.9	18.1	11.2	9.8	4.5	14.4
Manufacturing per cent	40.8	59.2	**100.0**	11.1	88.9	**100.0**	-50.8	150.8	**100.0**	47.6	52.4	**100.0**	19.8	80.2	**100.0**	37.2	62.8	**100.0**
Manufacturing *g*	0.78	1.13	**1.91**	0.30	2.43	**2.74**	-0.98	2.90	**1.93**	0.40	0.45	**0.85**	0.29	1.16	**1.45**	1.14	1.92	**3.07**

Table 5.5a Continued

	Japan 1982–97			Netherlands 1982–97			Norway 1976–97			Sweden 1973–97			United Kingdom 1970–97			United States 1970–97		
	STR	EFF	Total	STR	EFF	Total	STR	EFF	Total	STR	EFF	Total	STR	EFF	Total	STR	EFF	Total
Chemicals	8.4	1.1	9.4	-3.3	32.9	29.6	106.6	182.1	288.7	1.2	10.6	11.9	6.2	14.3	20.5	8.3	6.6	14.9
Food	-6.1	-7.5	-13.6	-2.8	15.3	12.5	-342.3	140.9	-201.4	-2.1	7.1	5.0	5.4	10.0	15.4	2.7	5.0	7.7
Iron and steel	-5.5	9.2	3.7	3.5	-3.5	0.0	-45.4	33.9	-11.5	-2.4	6.6	4.2	-3.1	3.4	0.3	-4.4	4.8	0.4
Machinery	45.2	52.6	97.9	17.8	10.4	28.3	-514.5	938.9	424.4	-22.5	80.7	58.2	-2.2	29.0	26.8	27.0	19.3	46.3
Transport equipment	12.4	-3.2	9.2	-3.4	5.7	2.3	-154.9	-171.6	-326.5	4.0	1.2	5.1	-3.3	13.2	9.9	-1.3	7.9	6.6
Non-ferrous metals	-0.5	0.6	0.1	-0.2	0.2	0.0	13.8	55.6	69.4	-0.1	1.7	1.6	0.0	0.6	0.6	0.1	0.9	1.0
Non-metallic minerals	-2.7	4.0	1.3	1.2	2.5	3.7	-126.6	80.6	-46.0	-5.3	3.4	-1.8	-3.7	5.8	2.1	2.3	-0.7	1.6
Paper	0.3	0.6	0.9	-3.1	13.8	10.7	90.6	22.7	113.3	3.1	7.1	10.2	2.9	8.6	11.5	3.3	6.2	9.4
Textiles	-17.6	6.2	-11.4	-1.5	0.3	-1.2	-117.8	39.0	-78.8	-5.0	0.8	-4.1	-2.7	3.5	0.8	0.7	4.3	5.1
Wood	--	--	--	-0.6	1.5	0.9	63.5	-106.1	-42.6	6.4	-5.1	1.4	--	--	--	--	--	--
Non-specified industry	-0.6	3.0	2.4	-37.0	50.2	13.2	68.9	-357.8	-288.9	11.0	-2.7	8.3	1.4	10.6	12.1	-3.9	10.9	7.0
Manufacturing per cent	33.3	66.7	**100.0**	-29.5	129.5	**100.0**	-958.1	858.1	**-100.0**	-11.6	111.6	**100.0**	1.0	99.0	**100.0**	34.8	65.2	**100.0**
Manufacturing *g*	0.57	1.14	**1.71**	-0.55	2.42	**1.87**	-0.74	0.67	**-0.08**	-0.25	2.44	**2.19**	0.02	2.26	**2.28**	1.11	2.08	**3.19**

Table 5.5b Percentage contribution of the structural effect (STR) and the efficiency effect (EFF) by sector to average annual growth rate (g) of aggregated labour productivity in manufacturing per country

	Belgium 1971–97			Denmark 1972–97			Finland 1971–97			France 1973–97			Germany (West) 1970–90			Italy 1970–97		
	STR	EFF	Total	STR	EFF	Total	STR	EFF	Total	STR	EFF	Total	STR	EFF	Total	STR	EFF	Total
Chemicals	7.8	13.2	20.9	6.4	14.3	20.7	2.0	4.1	6.1	2.1	13.7	15.8	1.8	11.6	13.4	-0.6	9.6	9.1
Food	3.0	11.5	14.5	-5.3	27.9	22.6	-0.9	7.9	7.1	5.8	9.7	15.5	-1.3	8.7	7.4	0.1	10.4	10.5
Iron and steel	-3.5	6.3	2.8	-0.7	2.3	1.6	0.5	3.9	4.5	-1.6	3.5	1.9	-4.2	6.5	2.3	-1.5	3.1	1.6
Machinery	-0.2	17.9	17.7	16.3	24.3	40.7	11.4	35.1	46.5	2.1	25.3	27.4	7.9	27.6	35.5	2.7	28.2	30.9
Transport equipment	4.0	6.3	10.3	-3.3	-2.6	-5.9	-2.2	5.5	3.2	0.1	9.0	9.1	11.7	8.2	19.9	-0.9	4.9	4.0
Non-ferrous metals	-1.2	4.2	3.0	--	--	--	0.3	1.2	1.5	-0.3	2.4	2.1	0.9	2.5	3.4	-0.7	1.3	0.7
Non-metallic minerals	-1.2	5.5	4.3	-5.3	1.3	-4.0	-0.7	3.3	2.5	-1.4	3.8	2.3	-2.4	5.2	2.8	-0.6	7.3	6.7
Paper	1.4	5.0	6.5	-1.6	6.6	5.0	-2.3	20.0	17.7	3.1	4.2	7.3	0.0	4.4	4.4	0.7	6.6	7.2
Textiles	-5.8	9.6	3.7	-8.8	6.8	-2.0	-4.6	4.2	-0.4	-6.8	6.0	-0.9	-8.3	6.9	-1.4	-0.1	15.0	15.0
Wood	0.1	0.9	0.9	1.1	1.7	2.7	-1.9	6.6	4.7	0.0	0.0	0.0	-0.8	1.7	0.9	-1.2	2.0	0.9
Non-specified industry	3.2	12.0	15.3	5.8	12.9	18.7	5.0	1.7	6.7	3.0	16.4	19.4	4.2	7.4	11.6	1.6	11.8	13.4
Manufacturing per cent	7.6	92.4	**100.0**	4.6	95.4	**100.0**	6.6	93.4	**100.0**	6.0	94.0	**100.0**	9.2	90.8	**100.0**	-0.3	100.	**100.0**
Manufacturing *g*	0.33	3.96	**4.29**	0.10	2.00	**2.09**	0.32	4.52	**4.84**	0.18	2.74	**2.92**	0.21	2.05	**2.26**	-0.01	3.94	**3.93**

Table 5.5b Continued

	Japan 1982–97			Netherlands 1982–97			Norway 1976–97			Sweden 1973–97			United Kingdom 1970–97			United States 1970–97		
	STR	EFF	Total	STR	EFF	Total	STR	EFF	Total	STR	EFF	Total	STR	EFF	Total	STR	EFF	Total
Chemicals	0.3	8.8	9.2	-1.8	25.9	24.1	0.2	20.9	21.0	4.6	6.0	10.6	0.4	18.3	18.7	1.0	15.9	16.8
Food	4.5	-7.1	-2.7	-2.0	16.5	14.6	10.9	-3.3	7.6	-0.3	6.8	6.5	1.2	13.9	15.1	-0.4	7.1	6.7
Iron and steel	-2.8	7.7	4.9	-1.5	2.6	1.1	-4.3	6.7	2.4	-2.5	6.6	4.0	-3.2	4.1	1.0	-4.5	3.8	-0.7
Machinery	4.0	66.7	70.6	2.8	24.7	27.5	10.0	35.5	45.5	4.5	45.4	49.8	2.1	24.8	26.9	0.8	52.2	53.0
Transport equipment	-0.7	10.1	9.4	-1.0	4.4	3.5	-12.9	8.1	-4.8	1.2	6.2	7.4	-2.6	13.0	10.3	1.4	4.0	5.4
Non-ferrous metals	-0.2	1.1	0.9	-0.7	1.1	0.5	-0.7	9.4	8.6	-0.4	1.9	1.6	-0.8	1.4	0.7	-0.3	1.0	0.7
Non-metallic minerals	-1.6	3.9	2.3	0.2	3.5	3.6	-2.1	4.2	2.0	-1.3	0.9	-0.4	0.0	2.4	2.4	-0.6	1.9	1.2
Paper	1.2	2.5	3.6	3.4	7.5	10.9	10.8	11.0	21.8	0.9	10.6	11.5	5.2	6.2	11.4	6.4	2.4	8.8
Textiles	-8.3	3.7	-4.6	-1.0	1.4	0.4	-5.8	4.5	-1.2	-4.1	2.2	-1.9	-4.2	6.1	1.9	-3.0	8.0	5.0
Wood	0.0	0.0	0.0	0.4	0.7	1.1	-3.6	6.9	3.4	-1.9	4.8	2.8	0.0	0.0	0.0	0.0	0.0	0.0
Non-specified industry	2.5	3.8	6.4	0.7	12.2	12.8	2.3	-8.6	-6.3	2.8	5.3	8.1	5.9	5.9	11.7	5.1	-2.0	3.1
Manufacturing per cent	-1.2	101.	**100.0**	-0.5	100.5	**100.0**	4.7	95.3	**100.0**	3.4	96.6	**100.0**	4.0	96.0	**100.0**	5.8	94.2	**100.0**
Manufacturing *g*	-0.04	3.94	**2.98**	-0.01	3.04	**3.03**	0.08	1.52	**1.60**	0.11	3.08	**3.19**	0.11	2.70	**2.82**	0.14	2.25	**2.39**

In Japan the structural changes towards a less energy-intensive manufacturing structure were mainly driven by a shift towards machinery and transport equipment, while energy shares decreased in textiles, foods, iron and steel and non-metallic minerals.

The negative structural effect in the Netherlands is the result of a shift in energy shares from non-specified industry, chemicals, transport equipment and paper towards machinery and iron and steel.

The major negative structural effect in Norway is mainly driven by a shift in energy shares towards chemicals and paper at the cost of decreasing energy shares in machinery, food, transport equipment and non-metallic minerals.

The negligible impact of structural changes in the United Kingdom is mainly due to the fact that a slight increase in energy shares in chemicals, food and paper outweighs a slight decrease in energy shares in non-metallic minerals, transport equipment and iron and steel.

In the USA a shift in energy share from iron and steel towards machinery and chemicals has been the main driving force behind the role of structural changes in improving aggregate manufacturing energy productivity improvement.

These findings confirm that in general a positive effect of total structural change on aggregate manufacturing energy productivity growth is to a large extent driven by a shift in energy shares from low-value added (energy-intensive) sectors – such as iron and steel, non-ferrous metals and non-metallic minerals – to higher value added (capital- and/or technology-intensive) industries – such as machinery, transport equipment, textiles and food – while the opposite is true in the case of an overall negative structural effect. Our results suggest however a few exceptions to this picture. For example Belgium realizes an overall positive effect of structural changes on aggregate manufacturing energy productivity growth in spite of a substantial decreasing energy share in the high value added machinery sector, while the same applies for France and Japan with respect to textiles. Moreover Belgium, West Germany, Japan and the USA combine an increasing energy share in the energy-intensive chemical sector with an overall positive structural effect, while the Netherlands combines a decreasing energy share in chemicals with an overall negative structural effect. A similar story is true for the paper sector: Belgium and France combine a substantial increase in energy share in the energy-intensive paper sector with an overall positive structural effect, while the opposite is true for the Netherlands, which realizes an overall negative structural effect in spite of a shift away from paper.

Of course these counter-examples can be explained from the simple fact that shifts in energy shares in one sector are sufficiently compensated for by shifts in other sectors. Moreover they might be due to data limitations, partly because in some countries (in particular Italy and the Netherlands) a significant role is played by the non-specified industry sector and partly because the two- and three-digit sector definitions that were used hide heterogeneity in production structure at the four-digit level.[18]

Finally, in Table 5.5b we present a similar breakdown of the total structural and efficiency effect as in Table 5.5a, but now for labour productivity. Table 5.5b shows that as for energy productivity, the aggregate manufacturing labour productivity improvements are to a large extent realised within the machinery sector. Moreover the table shows that although in all countries there is only a very limited impact of structural changes on aggregate manufacturing labour productivity growth, this does not imply that there were no changes in employment mix. It can be seen that the main structural change consisted of a decreasing employment share of the labour-intensive textiles sector (except for Italy) and an increasing employment share of the capital- and/or technology-intensive machinery sector (except for Belgium). Moreover in most countries this shift is accompanied by a shift in employment from iron and steel towards chemicals, food and paper.

Sectoral Biases in Productivity Growth Rates

For aggregate manufacturing we found in the previous section that although growth rates of labour productivity are substantially higher than growth rates of energy productivity, they nevertheless complement each other, and that the bias towards labour productivity growth is increasing over time. Do these conclusions also apply to the individual manufacturing sectors? To examine this issue we provide below some empirical evidence on the existence and development of potential sectoral biases towards either energy or labour productivity for each manufacturing sector.

For each manufacturing sector we calculated average annual growth rates of energy and labour productivity per country for the period 1970–97. They are presented in Figure 5.8 together with two regression lines through the origin, estimating the relationship between energy and labour productivity growth rates for the periods 1970–82 and 1982–97.[19] This leads to the following three conclusions. First, overall a positive correlation exists between energy and labour productivity growth rates, suggesting energy and labour productivity growth to be complements. There are however several exceptions. In several sectors, most notably in transport equipment, and paper and wood, several countries combine a positive labour productivity growth rate with a negative growth rate in energy productivity.

Second, also at this disaggregated level, labour productivity growth is in general higher than energy productivity growth, suggesting the existence of a bias towards labour-augmenting technological change.

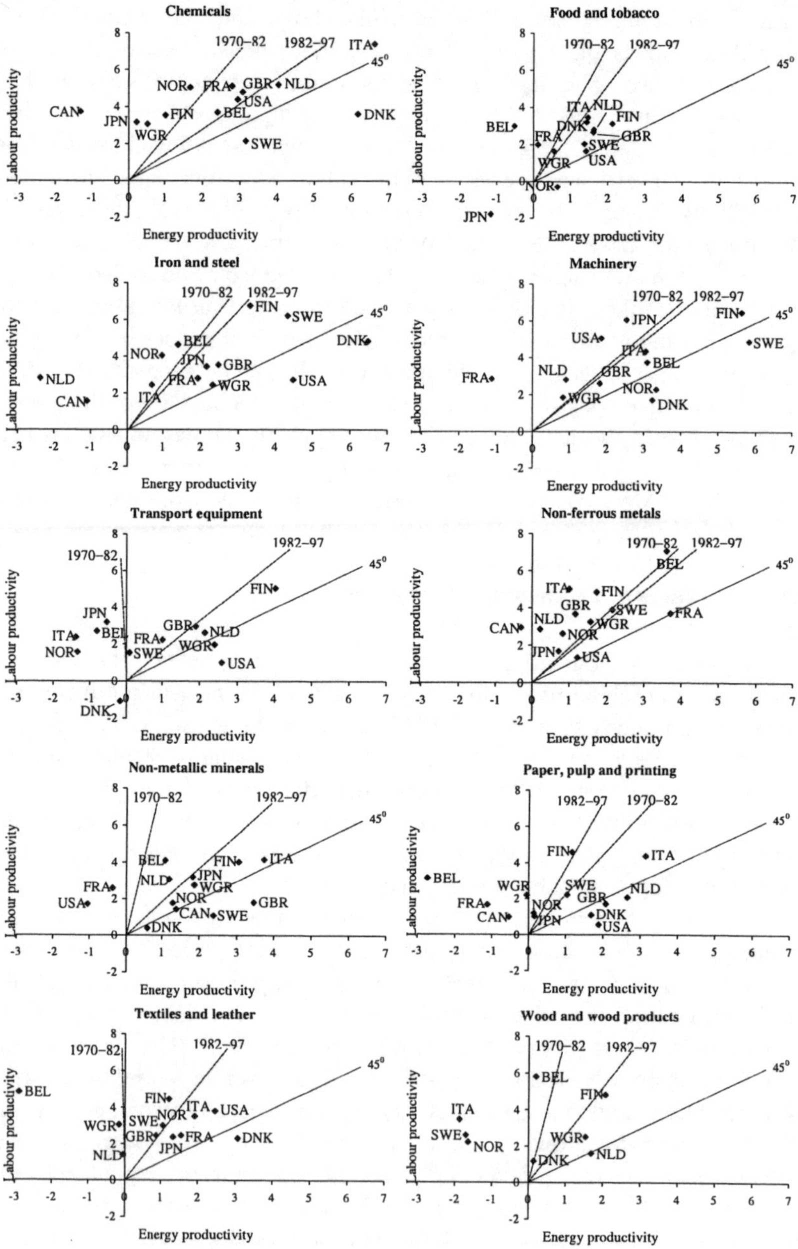

Figure 5.8. Energy and labour productivity manufacturing sectors, average annual growth rates

Third, over time this bias towards labour productivity growth is decreasing in all manufacturing sectors except for paper: in this sector the regression line for the period 1982–97 is steeper than for the period 1970–82. This result is in contrast with the increasing bias towards labour productivity growth which we found at the level of aggregate manufacturing and hence underlines the relevance of productivity analysis at a disaggregated level.

5.5 CONCLUSIONS

Technological change plays a crucial role in decoupling economic growth and environmental pressure. Technology-driven productivity growth is an important source of economic growth and plays an important role in realizing this decoupling, for example through increasing energy productivity. In this chapter we empirically examined the energy and labour productivity performance in 14 OECD countries over the last three decades. A principal aim of this chapter was to trace back macroeconomic productivity developments to developments at the level of individual sectors, in order to correct as much as possible for the impact of structural effects on productivity trends. Our analysis covered the period 1970–97 and distinguished 13 sectors, including ten manufacturing sectors, services, transport and agriculture. The research has been split into two parts, one focusing on the macroeconomic level and the other taking a closer look at the manufacturing sector.

At the macroeconomic level we found a diverse picture for trends in energy productivity, with substantial cross-country differences. Italy and Japan show a high energy productivity level while Canada, Finland, Norway and Sweden display a relatively low level of overall energy productivity. All other countries form a medium group. The USA tends to leave the lagging group over time to catch up with the medium group. At the level of aggregate manufacturing two groups of countries can be identified. Denmark, Italy, West Germany, France, Japan and the United Kingdom all show a relatively high energy productivity level, while the opposite holds for the USA, Netherlands, Belgium, Sweden, Australia, Finland, Canada and Norway. For labour productivity we found the well-known leading position for the USA, with other OECD countries showing a clear tendency to catch up at a macroeconomic level, a tendency that is less clear-cut at the manufacturing level.

A decomposition analysis revealed that, both at a macroeconomic level and at the manufacturing level, in most countries structural changes explain a substantial part of energy productivity growth rates, while they explain only a small part of labour productivity growth rates. At the macroeconomic level the dominating structural change consists of a shift in energy and employment shares from manufacturing towards services, while at the manufacturing level the positive structural effects are to a large extent driven by a shift of energy shares from low value added (energy-intensive) sectors (such as iron and steel, non-ferrous metals and non-metallic minerals) to higher value added (capital- and/or technology-intensive) industries (such as machinery, transport equipment, textiles and food) – while the opposite is true in case of an overall negative structural effect. Macroeconomic energy efficiency improvements are mainly realized within manufacturing, while for services the picture is highly diverse with a mix of positive and negative percentage contributions to aggregate energy efficiency improvements. In terms of labour productivity improvements, the main macroeconomic efficiency improvements are not only realized within manufacturing, but also within services. Within manufacturing, the machinery sector is the main source for both energy and labour productivity improvements. Finally, although the total structural effect on aggregate manufacturing labour productivity growth is small, there have been changes in employment mix: the main shifts include a decreasing employment share of the labour-intensive textiles sector(except for Italy) and an increasing employment share of the capital- and/or technology-intensive machinery sector (except for Belgium). Furthermore, in most countries this shift is accompanied by a shift in employment from iron and steel towards chemicals, food and paper.

An exploration of the relationship between energy and labour productivity growth rates revealed this relationship, with some exceptions, to be positive in most sectors, suggesting energy and labour productivity growth to be complements rather than substitutes. For most countries this conclusion holds also for services and transport. This may suggest that technological change is embodied in new capital goods which perform better than older capital goods in multiple dimensions, including a better performance in terms of both labour and energy productivity. This hypothesis assumes that knowledge is more or less a public good as a result of which the most recent capital goods embody state-of-the art technology in different dimensions. If this is true, firms and sectors investing in new capital goods in order to expand or replace existing production facilities, or to increase labour productivity, invest at the same time in energy-saving technological change. However more precise conclusions concerning these issues require a better insight into the nature of technological change through microeconomic research (see for example Newell et al. 1999), which is beyond the scope of this chapter.

Furthermore we found labour productivity growth rates in general to be substantially higher than energy productivity growth, while this bias towards labour productivity growth increased in aggregate manufacturing, transport and agriculture and decreased in services as well as in most manufacturing sectors. The latter result underlines the relevance of productivity analysis at a disaggregated level.

Finally, we found that, at several levels of aggregation, cross-country differences in energy productivity levels are substantially larger than cross-country differences in labour productivity levels. However our results suggest that whether these cross-country productivity differences tend to be decreasing or increasing over time depends on the level of aggregation. Therefore the next chapter is devoted to analysing the development of cross-country productivity differences over time. As part of this convergence analysis, in the next chapter we will also try to explain the productivity trends documented in this chapter, by searching for the main determinants of energy and labour productivity growth across countries.

APPENDIX 5.A

Table 5.A.1 Percentage shares of total energy consumption (E), employment (L) and GDP (Y) by sector, 1990

	Australia			Belgium			Canada			Denmark			Finland			France			Germany (W)		
	E	L	Y	E	L	Y	E	L	Y	E	L	Y	E	L	Y	E	L	Y	E	L	Y
Manufacturing	40	19	16	50	26	26	42	21	22	27	30	25	61	27	29	38	27	28	46	40	37
Services	7	63	67	13	56	57	18	60	61	17	45	53	5	42	44	17	50	57	16	42	50
Transport	48	5	6	35	6	8	36	5	5	48	8	9	27	7	7	41	5	5	37	5	4
Construction	2	7	7	--	9	7	1	8	9	2	10	8	1	12	12	1	10	7	--	9	7
Agriculture	3	6	4	2	3	2	3	5	3	7	8	6	6	12	8	3	8	4	2	4	2
Total	100	100	100	100	100	100	100	100	100	100	100	100	100	100	100	100	100	100	100	100	100

	Italy			Japan			Netherlands			Norway			Sweden			United Kingdom			United States		
	E	L	Y	E	L	Y	E	L	Y	E	L	Y	E	L	Y	E	L	Y	E	L	Y
Manufacturing	49	27	28	49	26	29	55	22	24	50	21	18	50	31	29	39	26	28	33	20	22
Services	4	46	55	14	47	50	5	57	59	15	51	54	17	46	51	12	55	54	16	66	66
Transport	43	6	5	31	6	7	30	6	5	31	9	16	31	7	6	47	7	7	50	4	4
Construction	--	9	8	2	11	11	1	9	7	2	10	7	--	10	10	1	9	9	--	7	5
Agriculture	4	12	4	4	10	3	9	6	5	2	9	5	2	6	4	1	3	2	1	3	2
Total	100	100	100	100	100	100	100	100	100	100	100	100	100	100	100	100	100	100	100	100	100

Table 5.A.2 Average annual growth rates of energy productivity (E) and labour productivity (L) in five sectors

		AUS	BEL	CAN	DNK	FIN	FRA	WGR	ITA	JPN	NLD	NOR	SWE	GBR	USA	OECD
		1970–97	1971–97	1970–97	1972–97	1971–97	1973–97	1970–90	1970–97	1982–97	1982–97	1976–97	1973–97	1970–97	1970–94	
MAN	E	0.41	1.91	0.45	2.74	1.93	0.85	1.45	3.07	1.71	1.87	-0.08	2.19	2.28	3.19	2.25
	L	1.06	4.29	1.79	2.09	4.84	2.92	2.26	3.93	2.98	3.03	1.60	3.19	2.82	2.39	2.69
		1974–96	1970–96	1973–97	1970–95	1970–96	1970–97	1970–90	1970–97	1982–96	1986–95	1976–95	1970–94	1970–96	1970–96	
TAS	E	1.18	-1.17	0.60	0.50	0.22	0.16	-0.33	0.16	-0.06	1.31	1.31	2.09	-0.15	1.02	0.39
	L	2.91	1.56	--	1.84	2.75	2.47	2.13	2.29	2.76	3.14	4.03	3.35	2.56†	1.11	2.03
SRV	E	-0.27	0.67	1.78	-1.05^	-2.82#	0.52°	2.43	-2.53	0.12	-2.56	0.76	2.18*	1.85	2.61	1.45
	L	0.64	0.96	0.49	1.63	2.87	1.54	2.16	0.97	2.12	0.18	0.48	1.63	0.22†	0.52	0.98
AGR	E	0.04	-2.58	-3.78	2.18	-1.16	0.36	1.50	-1.75	-3.42	-0.94	-4.07	0.58	2.19	2.68	0.18
	L	2.21	4.41	1.21	6.22	4.37	5.34	6.12	3.64	2.80	4.40	2.65	3.56	3.82	2.21	3.23
CST	E	-0.34	0.51	--	--	1.17	--	-1.59	-4.89	-0.37	-0.31	--	4.41	1.17	--	-0.42
	L	1.53	1.30	0.46	-0.32	2.23	1.49	1.41	0.74	1.43	0.07	--	2.70	--	-1.25	0.39

Notes:

*1986–1994 † 1970–1990 ^1972–1995 #1970–1995 °1985–1997.

The OECD average is weighted with each country's 1990 GDP share of total GDP per sector.

Table 5.A.3 Percentage shares of total manufacturing energy consumption (E), employment (L) and GDP (Y) by sector, 1990

	Australia			Belgium			Canada			Denmark			Finland			France			Germany (W)		
	E	L	Y	E	L	Y	E	L	Y	E	L	Y	E	L	Y	E	L	Y	E	L	Y
Chemicals	17	4	--	35	10	14	--	4	--	14	5	9	17	5	7	37	7	9	36	7	10
Food	12	12	--	6	14	15	--	12	--	27	17	21	4	12	12	9	13	13	6	9	11
Iron and steel	12	3	--	27	5	5	--	2	--	5	1	1	11	3	3	14	4	3	22	5	5
Machinery	2	19	--	3	24	21	--	20	--	8	34	30	3	26	27	8	29	26	4	37	34
Transport equipment	1	8	--	2	10	11	--	11	--	2	5	5	0	6	5	2	12	11	4	11	12
Non-ferrous metals	33	2	--	3	2	3	--	2	--	3	0	0	2	1	1	4	1	2	4	2	2
Non-metallic minerals	12	4	--	12	5	6	--	3	--	21	4	4	9	5	5	10	3	4	9	4	4
Paper	7	10	--	3	8	7	--	15	--	7	10	11	41	18	21	8	8	8	5	5	5
Textiles	2	8	--	2	12	7	--	9	--	3	7	5	1	8	4	3	10	6	2	6	4
Wood	1	5	--	0	2	1	--	5	--	4	3	2	6	8	7	2	2	2	1	2	1
Non-specified industry	0	25	--	7	8	10	--	16	--	4	13	12	8	9	9	3	11	16	7	11	12
Manufacturing	100	100	100	100	100	100	100	100	100	100	100	100	100	100	100	100	100	100	100	100	100

Table 5.A.3 Continued

	Italy			Japan			Netherlands			Norway			Sweden			United Kingdom			United States		
	E	L	Y	E	L	Y	E	L	Y	E	L	Y	E	L	Y	E	L	Y	E	L	Y
Chemicals	34	5	7	31	3	9	67	9	15	27	5	10	15	5	7	32	6	11	--	6	11
Food	5	8	10	4	10	11	8	17	16	6	18	16	4	9	10	10	11	13	--	9	11
Iron and steel	14	2	3	20	3	6	8	2	3	19	2	3	9	3	3	12	3	3	--	2	3
Machinery	7	26	27	6	36	38	3	31	26	3	25	25	6	35	31	10	32	26	--	27	27
Transport equipment	1	7	8	2	8	11	1	7	5	1	8	8	2	13	11	3	10	12	--	10	10
Non-ferrous metals	2	1	1	3	1	2	3	1	1	23	4	6	3	1	1	3	1	1	--	1	1
Non-metallic minerals	18	7	7	9	4	4	5	4	4	4	3	0	5	3	3	9	4	4	--	3	2
Paper	5	5	6	8	2	3	3	12	11	14	16	17	44	14	15	5	10	11	--	12	12
Textiles	5	23	16	3	7	2	1	6	3	0	4	2	1	3	2	3	10	6	--	10	5
Wood	0	3	2	--	3	1	0	2	2	3	6	5	6	6	6	0	2	1	--	4	3
Non-specified industry	9	13	13	14	22	13	0	9	14	0	10	7	4	8	9	12	12	11	--	15	15
Manufacturing	100	100	100	100	100	100	100	100	100	100	100	100	100	100	100	100	100	100	100	100	100

Table 5.A.4 Manufacturing sectors energy productivity average annual growth rates

	AUS	BEL	CAN	DNK	FIN	FRA	WGR	ITA	JPN	NLD	NOR	SWE	GBR	USA	OECD
	1970–97	1971–97	1970–97	1972–97	1971–97	1973–97	1970–90	1970–97	1982–97	1982–97	1976–97	1973–97	1970–97	1970–94	
Manufacturing	0.41	1.91	0.45	2.74	1.93	0.85	1.45	3.07	1.71	1.87	-0.08	2.19	2.28	3.19	2.25
Chemicals	--	2.4	-1.31	6.17	1.02	2.79	0.53	6.61	0.22	4.04	1.68	3.16	3.08	2.94	2.26
Food	--	-0.52	--	1.44	2.13	0.12	0.56	1.47	-1.16	1.63	0.65	1.38	1.62	1.42	0.64
Iron and steel	--	1.4	-1.09	6.48	3.32	1.93	2.32	0.69	2.47	-2.38	0.97	4.33	2.16	4.48	2.67
Machinery	--	3.09	--	3.23	5.62	-1.11	0.83	3.05	2.54	0.91	3.33	5.82	1.82	1.86	1.80
Transport equipment	--	-0.80	--	-0.17	4.04	1.01	2.40	-1.37	-0.53	2.14	-1.33	0.09	1.9	2.59	1.35
Non-ferrous metals	--	3.61	-0.3	--	1.74	3.72	1.58	1.01	0.72	0.22	0.83	2.17	1.17	1.23	1.31
Non-metallic minerals	--	1.09	1.39	0.59	3.05	-0.37	1.87	3.74	1.83	1.20	1.30	2.38	3.46	-1.05	1.30
Paper	--	-2.73	-0.54	1.69	1.17	-1.12	-0.05	3.14	0.16	2.66	0.14	1.04	2.08	1.89	1.31
Textiles	--	-2.90	--	3.06	1.23	1.56	-0.16	1.92	1.34	-0.05	1.06	1.07	0.87	2.46	1.59
Wood	--	0.2	--	0.14	2.07	--	1.54	-1.87	--	1.69	-1.64	-1.70	--	--	-0.01

Note:

The OECD average is weighted with each country's 1990 GDP share of total GDP per sector.

Table 5.A.5 Manufacturing sectors labour productivity average annual growth rates

	AUS	BEL	CAN	DNK	FIN	FRA	WGR	ITA	JPN	NLD	NOR	SWE	GBR	USA	OECD
	1970–97	1971–97	1970–97	1972–97	1971–97	1973–97	1970–90	1970–97	1982–97	1982–97	1976–97	1973–97	1970–97	1970–94	
Manufacturing	1.06	4.29	1.79	2.09	4.84	2.92	2.26	3.93	2.98	3.03	1.60	3.19	2.82	2.39	2.69
Chemicals	--	3.70	3.74	3.61	3.55	5.12	3.06	7.41	3.16	5.19	5.04	2.16	4.80	4.40	4.22
Food	--	2.99	1.34	3.21	3.12	1.99	1.63	3.46	-1.79	2.81	-0.31	2.04	2.67	1.65	1.25
Iron and steel	--	4.63	1.56	4.83	6.77	2.80	2.44	2.44	3.53	2.82	4.05	6.22	3.44	2.72	3.03
Machinery	--	3.74	1.90	1.70	6.47	2.86	1.85	4.34	6.07	2.81	2.28	4.87	2.61	5.05	4.40
Transport equipment	--	2.70	2.54	-1.08	5.05	2.23	1.97	2.37	3.19	2.61	1.58	1.53	2.96	1.00	2.02
Non-ferrous metals	--	7.07	2.95	4.28	4.85	3.70	3.26	5.03	1.68	2.86	2.62	3.89	3.66	1.35	2.41
Non-metallic minerals	--	4.10	1.43	0.39	4.02	2.58	2.76	4.15	3.21	3.08	1.77	1.08	1.80	1.71	2.71
Paper	--	3.14	1.00	1.09	4.60	1.66	2.20	4.37	1.02	2.07	1.19	2.22	1.71	0.56	1.27
Textiles	--	4.84	2.73	2.32	4.46	2.47	3.06	3.52	2.38	1.41	3.02	3.00	2.47	3.81	3.28
Wood	--	5.81	1.98	1.17	4.82	4.10	2.52	3.46	0.31	1.60	2.22	2.60	0.66	0.95	1.53

Note:

The OECD average is weighted with each country's 1990 GDP share of total GDP per sector

APPENDIX 5.B

Essentially, one can use two types of indicator to measure energy productivity, each measuring economic activity (output or production) in a different way. An economic indicator measures economic activity in monetary values, while a physical indicator measures economic activity in terms of physical production. The most common unit of monetary value is value added (GDP)[20] while physical production is usually expressed in terms of physical volume of production (weight and number of products). Both types of indicator have their advantages and disadvantages (see for example Phylipsen et al. 1997, 1998). The main advantage of using a physical indicator, or so-called Specific Energy Consumption (SEC), is that it measures a direct relationship between the volume of production and energy consumption (for example MJ/tonne). An economic indicator does not measure such a direct relationship, since it measures not only SEC but also changes in the mix and characteristics of products and feedstock as well as changes in market-based product prices. In addition since physical indicators are necessarily developed at a lower level of aggregation than economic indicators – because of the use of physical units of production – the influence of the structure of output on the aggregate energy productivity performance is by definition of less importance. But physical indicators also have their problems. The three most important disadvantages are inherent difficulties of aggregation (how to add up different levels of energy services in physical terms), lack of useful physical indicators of economic activity (in particular in the energy-extensive sectors) and limited data availability. Especially in sectors with a large variety of products and a large degree of processing, using physical indicators requires a large amount of data.

Although not many systematic comparisons between physical and economic indicators have been made so far, there is some evidence of substantial differences between the two indicators at the sector level, especially in the short run (Farla and Blok 2000, Freeman et al. 1997, Worrell et al. 1997). It is to be noted however that in general a value added-based energy intensity seems to follow the SEC better than other economic indicators (Worrell et al. 1997). An important criterion in choosing between the two different approaches is the research question at hand. If one is primarily interested in the relationship between energy consumption and volume of production at the process level, one may not want to use economic indicators since they do not always adequately capture physical developments at such a micro level. If one is primarily interested in the relationship between energy use and economic growth (measured in terms of GDP), an economic indicator might be most suitable to examine energy productivity developments.

We have chosen in this study to use an economic indicator, measuring energy productivity by gross value added per unit of final energy consumption. The main reason is that lack of physical production data would prevent us from conducting a systematic cross-country analysis for a broad range of sectors. Moreover we adopt a macroeconomic view, examining the role of two production factors in driving aggregate productivity developments. In addition since we compare trends in energy and labour productivity, using gross value added to measure economic activity establishes a link with the existing empirical literature focusing on labour and total factor productivity developments. What is more, our disaggregated level of analysis includes sufficient sectoral detail to account for the main part of structural changes on aggregate productivity growth, and hence provides a reasonable indication of energy efficiency developments. Furthermore, the latter is also true because we perform a long-term analysis, which is not so much biased by short-run fluctuations in value added figures. Finally, our analysis does not include developing countries and centrally planned economies, limiting measurement errors for the value added indicator due to black, grey or missing markets.

APPENDIX 5.C

This appendix provides technical details and a brief discussion of alternative decomposition methods and their relation to one another. Moreover we will provide the motivation for the choice for the decomposition method used in this chapter. Index number decomposition analysis is a methodology to decompose changes in an aggregate indicator into contributions from several specified factors. In this chapter we decompose changes in the aggregate energy and labour productivity into a contribution from an ‘efficiency effect’ and a contribution from a ‘structural effect’. The efficiency effect captures the net effect of changes in sectoral energy or labour productivity on the change in aggregate energy or labour productivity, holding the sector mix constant. The structural effect captures the net effect of changes in sector mix on the change in aggregate energy or labour productivity, holding sectoral energy or labour productivity levels constant. Therefore this methodology is sometimes referred to as shift-share analysis.

In the context of decomposing aggregate energy and labour productivity the methodology is based on the following definitions:

$$\frac{Y_t}{E_t} = \sum_i \frac{Y_{i,t}}{E_{i,t}} \frac{E_{i,t}}{E_t} \tag{5.C.1}$$

$$\frac{Y_t}{L_t} = \sum_i \frac{Y_{i,t}}{L_{i,t}} \frac{L_{i,t}}{L_t} \tag{5.C.2}$$

with $Y_{i,t}$, $E_{i,t}$ and $L_{i,t}$ being respectively GDP, final energy consumption and total employment of sector *i*. Similar, Y_t, E_t and L_t are respectively aggregate GDP, aggregate final energy consumption and aggregate total employment. So equation (5.C.1) says that the aggregate energy productivity is the sum of each sector's energy productivity level (the first term at RHS) multiplied by its energy share (the second term at RHS). Equation (5.C.2) defines the same relationship in terms of labour productivity.

For convenience we define $I_E = \frac{Y}{E}, I_L = \frac{Y}{L}$ and $S_{Ei} = \frac{E_i}{E}, S_{Li} = \frac{L_i}{L}$, such that (5.C.1) and (5.C.2) can be summarized by

$$I_{p,t} = \sum_i I_{pi,t} S_{pi,t} \quad \text{with } p = E, L \tag{5.C.3}$$

which says that the aggregate productivity index is the product of the sum of each sector's energy productivity level multiplied by its factor share. For the actual decomposition of aggregate productivity I into an efficiency index I_i and a structural index S_i several methods and functional forms can be used, requiring a choice with respect to four issues (Ang 1995a).

1. **Additive or multiplicative technique**
An additive technique builds upon the equation $\Delta I = \Delta I_i + \Delta S_i$ with $\Delta I = I_T - I_0$, $\Delta I_i = I_{i,T} - I_{i,0}$ and $\Delta S_i = S_{i,T} - S_{i,0}$. A multiplicative technique builds upon the equation $\Delta I = \Delta I_i \cdot \Delta S_i$ with $\Delta I = I_T / I_0$, $\Delta I_i = I_{i,T} / I_{i,0}$ and $\Delta S_i = S_{i,T} / S_{i,0}$.

2. **The decomposition method**
To actually calculate the Efficiency Effect ΔI_i and the Structural Effect ΔS_i, essentially three methods exist: (1) the General Parametric Divisia Method 1, (2) the General Parametric Divisia Method 2 and (3) the Refined Divisia Method. An important distinguishing feature of the RDM is that it leaves no residual term; that is, there is no part of the change in aggregate change left as unexplained. Below, we present the three methods (for energy productivity), applying the additive technique

and with 0 as the initial year and T as the final year of the decomposition period:

a. General Parametric Divisia Method 1 (PDM 1)

$$\Delta I_i = \sum \left\{ \frac{Y_0}{E_{i,o}} + \alpha \left(\frac{Y_T}{E_{i,T}} - \frac{Y_0}{E_{i,o}} \right) \right\} * \ln \left(\frac{I_{i,T}}{I_{i,o}} \right)$$

$$\Delta S_i = \sum \left\{ \frac{Y_0}{E_{i,o}} + \beta \left(\frac{Y_T}{E_{i,T}} - \frac{Y_0}{E_{i,o}} \right) \right\} * \ln \left(\frac{S_{i,T}}{S_{i,o}} \right)$$

b. General Parametric Divisia Method 2 (PDM 2)

$$\Delta I_i = \sum \left\{ S_{i,o} + \alpha \left(S_{i,T} - S_{i,o} \right) \right\} * \left(I_{i,T} - I_{i,o} \right)$$

$$\Delta S_i = \sum \left\{ I_{i,o} + \beta \left(I_{i,T} - I_{i,o} \right) \right\} * \left(S_{i,T} - S_{i,o} \right)$$

c. Refined Divisia Method (RDM)

$$\Delta I_i = \sum \left\{ \frac{\left(\frac{Y_T}{E_{i,T}} - \frac{Y_0}{E_{i,o}} \right)}{\ln \left(\left(\frac{Y_T}{E_{i,T}} \right) / \left(\frac{Y_0}{E_{i,o}} \right) \right)} \right\} * \ln \left(\frac{I_{i,T}}{I_{i,o}} \right)$$

$$\Delta S_i = \sum \left\{ \frac{\left(\frac{Y_T}{E_{i,T}} - \frac{Y_0}{E_{i,o}} \right)}{\ln \left(\left(\frac{Y_T}{E_{i,T}} \right) / \left(\frac{Y_0}{E_{i,o}} \right) \right)} \right\} * \ln \left(\frac{S_{i,T}}{S_{i,o}} \right)$$

3. **The functional form**

For the RDM no additional specification has to be made. The exact form of the decomposition method in case of PDM1 and PDM2 however depends on the parameter values chosen. The choice of the parameter values implies giving weight to the start and end year of the decomposition period. Three parameter values are most widely used, specifying three decomposition methods:

a. Laspeyres index ($\alpha = \beta = 0$), giving all weight to base year *0*:

$$\text{PDM1 } \Delta I_i = \sum \left\{ \frac{Y_0}{E_{i,0}} \right\} * \ln\left(\frac{I_{i,T}}{I_{i,0}} \right)$$

$$\text{PDM1 } \Delta S_i = \sum \left\{ \frac{Y_0}{E_{i,0}} \right\} * \ln\left(\frac{S_{i,T}}{S_{i,0}} \right)$$

$$\text{PDM2 } \Delta I_i = \sum S_{i,0} * \left(I_{i,T} - I_{i,0} \right)$$

$$\text{PDM2 } \Delta S_i = \sum I_{i,0} * \left(S_{i,T} - S_{i,0} \right)$$

b. Marshall-Edgeworth of Divisia-Törnqvist ($\alpha = \beta = 0.5$), giving equal weight to base 0 and end year T:

$$\text{PDM1 } \Delta I_i = \sum \left\{ 0.5 * \left(\frac{Y_0}{E_{i,0}} + \frac{Y_T}{E_{i,T}} \right) \right\} * \ln\left(\frac{I_{i,T}}{I_{i,0}} \right)$$

$$\text{PDM1 } \Delta S_i = \sum \left\{ 0.5 * \left(\frac{Y_0}{E_{i,0}} + \frac{Y_T}{E_{i,T}} \right) \right\} * \ln\left(\frac{S_{i,T}}{S_{i,0}} \right)$$

$$\text{PDM2 } \Delta I_i = \sum \left\{ 0.5 * \left(S_{i,0} + S_{i,T} \right) \right\} * \left(I_{i,T} - I_{i,0} \right)$$

$$\text{PDM2 } \Delta S_i = \sum \left\{ 0.5 * \left(I_{i,0} + I_{i,T} \right) \right\} * \left(S_{i,T} - S_{i,0} \right)$$

c. Paasche index ($\alpha = \beta = 1$), giving all weight to end year *T:*

$$\text{PDM1 } \Delta I_i = \sum \left\{ \frac{Y_T}{E_{i,T}} \right\} * \ln\left(\frac{I_{i,T}}{I_{i,0}} \right)$$

$$\text{PDM1 } \Delta S_i = \sum \left\{ \frac{Y_T}{E_{i,T}} \right\} * \ln\left(\frac{S_{i,T}}{S_{i,0}} \right)$$

$$\text{PDM2 } \Delta I_i = \sum S_{i,T} * \left(I_{i,T} - I_{i,0} \right)$$

$$\text{PDM2 } \Delta S_i = \sum I_{i,T} * \left(S_{i,T} - S_{i,0} \right)$$

Alternatively it has been proposed that the parameter values are made 'endogenous' by equating the formula of PDM1 to that of PDM2 for each estimated effect, a method referred to as the Adaptive Weighting Parametric Divisia Method (AWT-PDM). This 'smoothing' process makes the decomposition results independent of a somewhat arbitrary choice for PDM1 or PDM2.

4. **A period-wise or a time-series analysis**
   A time-series approach uses yearly data to define base year 0 and end year *T* while a period-wise approach uses data for the first and the last year of a specified time period only.

In this study we have chosen a time-series analysis, the additive technique and the Refined Divisia Method. The latter implies that no additional choice needs to be made with respect to the parameter values α and β. We have chosen to use a time-series approach because it yields more insight into energy productivity developments over subsequent years – and our database contains yearly data. Moreover the decomposition results given by time-series analysis are less dependent on the decomposition method used, as compared to period-wise decomposition. We have chosen to use the additive technique because we are interested in decomposing the absolute change in energy and labour productivity, rather than a relative change. Finally we have chosen to use the RDM because this method gives, contrary to the other methods, perfect decomposition irrespective of the pattern exhibited by the data and leaving no residual term. Moreover this method has the advantage that it can handle zero values in the data set effectively, while the other methods cannot. For further details on decomposition methodology and a systematic survey in energy studies we refer to Ang (1995a, 1999) and Ang and Zhang (2000).

NOTES

1. Alternatively, one can also make use of physical productivity indicators to measure energy and labour productivity. We refer to Appendix 5.B for a brief discussion on the use of physical versus economic indicators of output, and hence productivity.
2. Hence we do not analyse explicitly the impact of changes in fuel mix on overall energy efficiency improvements.
3. This limited availability is due to some problems inherent in the industry-of-origin approach: producer prices (that is, production values divided by output quantities) may not properly account for cross-country quality differences and imply aggregation problems because they are available only for a sample of goods (partly because of confidentiality problems), and because the production structure among countries tends to be less comparable than the consumption structure due to specialization tendencies in production according to comparative advantage (Pilat 1996).
4. Note that most studies analysing energy efficiency developments use energy intensity as an indicator, being the inverse of energy productivity. We prefer to use energy productivity simply because it establishes a direct link with the empirical literature on labour productivity developments.
5. Hence in this chapter 'macroeconomic' refers to the sum of the manufacturing, transport, services and agriculture sectors and thus excludes construction, households and the energy production sector.
6. Although we have some reason to believe that this result might be due to poor data (see next section) the relatively good energy productivity performance of Italy in services is also found by Schipper and Meyers (1992: 185) who document for Italy in 1973 and 1988 an energy intensity level in services that is substantially lower than in eight other OECD countries.
7. The SD of the log of productivity (y) measures variation across countries i according to:

$$\sqrt{\frac{1}{n}\sum_{i=1}^{n}\left(\log y_i - \overline{\log y}\right)^2}, \ \overline{\log y} = \frac{1}{n}\sum_{i=1}^{n}\log y_i$$

   For further discussion see Chapter 6 (section 6.3).
8. For early applications of this methodology to measure the impact of technological change and changes in labour and/or capital shares on aggregate (total factor productivity) growth, see for example Maddison (1952) and Massell (1961). For recent applications, including cross-country comparisons, see Dollar and Wolff (1993), van Ark (1996) and Fagerberg (2000). Cross-country decomposition analyses of energy use can be found for example in Morovic et al. (1987, 1989), Greening et al. (1997), Howarth et al. (1991), Schipper and Meyers (1992), Park et al. (1993), Eichhammer and Mannsbart (1997) and Unander et al. (1999). From these studies, only van Ark (1996) and Schipper and Meyers (1992) include non-manufacturing sectors, while with the exception of Fagerberg (2000) and Park et al. (1993) all other studies focus on OECD countries.
9. A closer look at the data reveals that this result is due to an exceptionally low initial level of energy consumption in services in these countries, which then increases relatively fast over time to converge to an average level. Since we have no breakdown of energy data for the underlying subsectors we cannot explore this issue any further, but it might just be due to poor quality of the data. See also Ramirez et al. (2002) who found, in a detailed analysis of the Dutch service sector for the period 1984–98, a minor increase of energy productivity, which has been hardly affected by structural changes.
10. Note that the exact period differs per country due to data restrictions. We refer to Table 5.A.2 in Appendix 5.A for an overview of the periods used for each country as well as the sectoral growth rates per country (the same as in Figure 5.4 but in table format).
11. These percentages are in line with IEA data of total OECD in 1997: chemicals: 39 per cent; iron and steel: 11 per cent ; paper: 10 per cent ; non-metallic minerals: 9 per cent.
12. Note that manufacturing sector shares of value added are calculated for the sum of 12 OECD countries (excluding Australia and Canada due to lack of data).
13. For the same sample of countries but using energy consumption data from partly different

sources, Unander et al. (1999) distinguish three groups of countries for manufacturing energy intensity, which differs slightly from our picture in classifying the Netherlands, Sweden and the USA in a medium group.

14. Note that we do not take the USA as the reference country because it lacks a sectoral breakdown of oil and natural gas consumption at this level of disaggregation (see section 5.2).
15. Due to limited data availability, Australia and Canada are excluded from the analysis. Moreover for the same reason in France, Japan, the United Kingdom and USA the wood sector and in Denmark the non-ferrous metals sector are excluded from the decomposition analysis.
16. The results also depend to some extent on the level of aggregation. As noted in section 5.2, the higher the level of disaggregation the better, but even more important is an adequate distinction between factor-intensive and factor-extensive sectors in order to reduce the likelihood of efficiency performance figures being biased by the impact of intra-sectoral structural changes. However, in a European cross-country decomposition analysis of energy efficiency in the manufacturing industry, Eichhammer and Mannsbart (1997) concluded that, apart from data-related methodological problems, an analysis at a two-digit level suffices to isolate the main structural effects on aggregate productivity developments. After disaggregating several energy-intensive sectors to a four-digit level, they found intra-sectoral structural changes to be responsible for at maximum ten per cent of the observed aggregate energy-intensity changes.
17. Note that in France, Norway and in particular the Netherlands a substantial part of the efficiency improvement is realized within the non-specified industry sector (NSI). The same holds for structural changes in Italy, Norway and, again particularly, the Netherlands. NSI contains rubber (355) and plastic products (not classified elsewhere) (356), furniture (332) and professional, scientific, measuring and controlling equipment (not classified elsewhere), photographic and optical goods (385). Furthermore, it contains energy consumption for which no sectoral breakdown can be given. Whereas NSI is rather unimportant in most countries with an average share of 2 per cent of total energy consumption, in the countries mentioned before the share of NSI in manufacturing GDP is on average about 12 per cent. In sum one should read the results with caution since an efficiency improvement and a changing energy share in NSI is partly due to developments in the above-mentioned sectors (ISIC 355, 356, 332 and 385) and might be partly due to data inaccuracy.
18. Recall that the chemicals sector is built up from the energy-intensive industrial chemicals subsector (ISIC 351) and the energy-extensive other chemical products subsector (ISIC 352). Similarly, the paper sector is built up from the energy-intensive paper and pulp sector (ISIC 341) and the energy-extensive printing sector (ISIC 342. Hence the observed shifts in energy shares might be characterized as intra-sectoral shifts (see note 16). For example it is known that in the Netherlands the share of industrial chemicals in the chemical industry has been substantially reduced over time (CPB 2000: 63–68). As noted before, until consistent and internationally comparable energy and economic data become available for a more detailed breakdown of these sectors, the decomposition results should be interpreted with caution.
19. Note that the exact period differs for each country due to data restrictions. We refer to Tables 5.A.4 and 5.A.5 in Appendix A for an overview of the periods used per country, the sectoral growth rates per country (the same as in Figure 5.8 but in table format) as well as the weighted average sectoral growth rates for the sum of the OECD countries included in this study.
20. Alternative, and less common, value-based measures for economic activity are gross output, value of shipments and value of production. Apart from the fact that these measures are not reported in the ISDB or STAN databases, value added is an appropriate measure of economic activity because it measures incremental value added by sector and thus avoids double counting of production.

6. Sectoral energy and labour productivity convergence

6.1 INTRODUCTION

Economies differ, and so does productivity performance. Of course economies also interact, and hence productivity developments are thought to be determined not only by developments within a particular country or sector, but also by what is happening in the outside world. Therefore a key issue in understanding long-run productivity performance is whether the process of economic growth tends to involve reductions in productivity differences among countries, for example due to capital accumulation or technology transfers. In Chapter 5 we found evidence of substantial differences in energy and labour productivity performance across countries and across sectors. A decomposition analysis at different levels of aggregation showed that after correcting for the impact of structural changes, there remain substantial technology-driven productivity growth differentials among OECD countries. Contrary to the previous chapter with its emphasis on the relative performance of different countries, we focus in this chapter on the group of countries as a whole and ask whether the observed cross-country productivity differences are persistent or whether they tend to decline over time. By searching for the determinants of (differences in) energy- and labour productivity growth across countries and across sectors, this chapter provides an explanatory analysis of sectoral trends in energy and labour productivity performance across countries, in addition to the descriptive analysis of Chapter 5.

The introduction of new or endogenous growth theories, as discussed in Chapter 2, has generated renewed interest in this question and has induced a wave of empirical research on cross-country convergence of per capita income as well as of labour productivity or total factor productivity (for example Barro and Sala-i-Martin 1992, Mankiw et al. 1992, Islam 1995). This issue obviously has important implications for the international welfare distribution. At the same time – as argued in Chapter 5 – the need for sustainable development implies that not only labour productivity but also energy productivity have to increase. Consequently important questions arise concerning the dynamics of cross-country differences of energy and labour productivity. Are these differences decreasing, or is the gap between leading and backward countries getting larger? Does energy productivity convergence follow patterns of labour productivity convergence? Do relatively inefficient countries catch up with technological 'leaders'? And if so, how quickly and by what means? Do convergence patterns differ substantially across sectors? We will answer these questions by simultaneously carrying out an empirical analysis of cross-country energy and labour productivity convergence at a detailed sector level, using a new dataset that merges energy data and economic data for 13 sectors and 14 OECD countries, covering the period 1970–97.

In several respects the research presented in this chapter differs from previous empirical research on cross-country convergence. By including energy productivity developments, our analysis differs from the empirical macroeconomic convergence literature that focuses on convergence of per capita income, labour productivity and total factor productivity. Moreover in spite of many existing cross-country studies on energy productivity or energy intensity developments and their determinants (for example Howarth et al., 1991; Miketa, 2001; Schipper and Meyers, 1992; Unander et al., 1999), to the best of our knowledge this literature lacks a systematic convergence analysis from a macroeconomic perspective. Hence we add to the existing literature a systematic comparison of energy and labour productivity convergence. Finally, we do so at a detailed sectoral level. By looking at cross-country convergence patterns within sectors, our analysis differs from virtually all convergence studies in the empirical growth literature, since they employ aggregated data. Important exceptions are sectoral studies by Dollar and Wolff (1988, 1993) and Bernard and Jones (1996a, b) who – using partly the same data source (OECD's ISDB) – conclude that a convergence analysis of aggregate productivity levels masks substantial differences at the sectoral level. We drew a similar conclusion in the previous chapter, and hence an important goal of this chapter is to examine and compare patterns of productivity convergence at different levels of sectoral aggregation. Our sectoral approach corrects for most of the impact of changes in the structure of production on aggregate productivity developments, and hence establishes a closer link to issues concerning international convergence of technology-driven productivity performance. Our analysis differs from the previously mentioned sectoral convergence analyses in comparing labour and energy productivity convergence, in further disaggregating the manufacturing sector into ten subsectors,[1] in using more recent data, and in carrying out a more extensive search for country- and sector-specific factors to explain productivity convergence patterns.

The chapter proceeds as follows. Section 6.2 contains a brief review of the convergence issue in the economic literature, in order to provide some theoretical background to our empirical analysis and to introduce several concepts used throughout the chapter. Section 6.3 describes the data used. In section 6.4 we analyse the development of cross-country differences of energy and labour productivity levels within sectors over time. In section 6.5 we use a panel data approach to test the proposition that sectoral growth rates of energy and labour productivity are inversely related to their initial levels of energy and labour productivity, indicating possible patterns of catching up. In addition we try to identify the country- and sector-specific fundamentals determining differences in energy and labour productivity developments. Section 6.6 summarizes and concludes.

6.2 THE ISSUE OF CONVERGENCE

The concept of productivity convergence has its roots in traditional neoclassical growth theory, with its central notion of a transitional growth path to a steady state. The Solow–Swan model (Solow 1956, Swan 1956) postulates convergence of per capita income, driven by the assumption of diminishing returns to capital accumulation at the economy-wide level. The dynamics of the model imply that initial differences in per capita income and capital endowments will vanish in the long run, due to declining growth rates as countries approach their steady state. In the steady state, diminishing returns are offset by technological progress, the principal source of long-run economic growth. The new or endogenous growth theory (Lucas 1988; Romer 1986, 1990) yields a more diverse picture concerning patterns of convergence. In this view economic growth is ultimately driven by accumulation of knowledge or human capital, which is (at least partially) a public good. Hence cross-country convergence depends on the extent of international knowledge spillovers, allowing less productive countries to catch up with more advanced economies. As such, endogenous growth theory supports the old hypothesis of the existence of an 'advantage of backwardness' (Gerschenkron 1952), suggesting that being relatively backward in productivity carries a potential for rapid advance (see for example Abramovitz 1986). At the same time, endogenous growth theory suggests that growth differentials may persist or even increase: learning effects, externalities and market imperfections allow for economy-wide increasing returns to capital accumulation and the existence of multiple steady-states. A mixed view on convergence patterns also emerges if one takes into account the role of international trade: on the one hand trade will enhance cross-country convergence through knowledge diffusion and increasing competition, but on the other hand it may contribute to cross-country divergence since trade advances international specialization (for example Grossman and Helpman 1991).

These various approaches generated some degree of controversy around the issue of convergence and caused the convergence hypothesis to be the subject of extensive empirical research (see Islam 2003 for a recent survey). The stage for this convergence debate was set by Baumol (1986), who reported a strong negative relationship between the initial level of labour productivity and its subsequent growth over a long period (1870–1979), which he argued to be strong evidence in favour of convergence. Abramovitz (1986) presented similar evidence, arguing that catch-up growth has been most prominent in the period since 1945.

This position was challenged by DeLong (1988) who argued that Baumol's results suffered from a sample bias, in that his analysis has been confined to a sample of countries that have become rich and developed; if one takes a sample of countries that in 1870 seemed likely to converge, the evidence of convergence is less clear-cut. In addition a number of studies have presented evidence of income convergence across countries, by explicitly testing (augmented versions of) the Solow growth model (for example Barro 1991, Barro and Sala-i-Martin 1992, Mankiw et al. 1992, Islam 1995). These empirical cross-country growth analyses raised the important question of whether countries converge to a global or rather to a local steady state, the latter implying that convergence is conditional on cross-country differences in steady-state characteristics. They are, respectively, referred to as unconditional (or absolute) and conditional (or relative) convergence. This idea has been formalized by Durlauf and Johnson (1992) and confirmed by several studies in this field, some of them suggesting the existence of convergence clubs: groups of countries converging to different steady states (for example Chatterji 1992, Chatterji et al. 1993, Quah 1997).[2]

From this literature it follows that convergence can be understood in terms of levels and growth rates, which translates into a distinction between so-called σ-convergence and β-convergence (for example Barro 1991, Barro and Sala-i-Martin 1992). The former refers to a decreasing variance of cross-country differences in productivity levels, while the latter suggests a tendency of countries with relatively low initial productivity levels to grow relatively fast, building upon the proposition that growth rates tend to decline as countries approach their steady state. Obviously, σ-convergence and β-convergence are closely related. A narrowing dispersion of cross-country productivity differences implies that countries with a relatively poor initial productivity performance tend to grow relatively fast. However as has been argued by Quah (1993), a statistically significant inverse relationship between the initial level and the growth rate of productivity performance can be consistent with constant or even increasing cross-country productivity differences – a phenomenon known as Galton's Fallacy of regression towards the mean.[3] In other words, β-convergence is a necessary but not a sufficient condition for σ-convergence. In this chapter we will explore both patterns of σ-convergence and β-convergence. Moreover we test whether convergence is conditional or unconditional and which are the factors explaining cross-country differences in labour and energy productivity growth.

6.3 DATA

We employ the same dataset as in the previous chapter, including 13 sectors, 14 OECD countries and covering the period 1970–97. Compared to Chapter 5, in this chapter we include several additional variables: investment, energy prices, compensation of employees, exports and imports – all at the sectoral level. The sector-specific energy prices are constructed by dividing sector-specific expenditures on energy over total sectoral energy consumption. The sector-specific expenditures are calculated as the product of the sectoral consumption of the four main energy carriers (coal, natural gas, electricity, oil), available from the IEA Energy Balances, and the annual price of each energy carrier at the aggregate industrial sector, available from the IEA Energy Prices and Taxes series. In addition some missing aggregate energy price data series have been constructed. Again all currency-denominated variables are in 1990 US dollars and have been converted by the OECD using 1990 purchasing power parities.[4] Finally, we measure energy and labour productivity again by, respectively, value added per ktoe energy consumption and value added per worker in full-time equivalents. For further details on data and sector classification we refer to Chapter 5 and Mulder (2003).

6.4 σ-CONVERGENCE

This section deals with the notion of convergence in terms of levels. Do the cross-country differences in energy and labour productivity levels that we observed in Chapter 5 decrease over time? Are patterns of energy productivity convergence similar to those of labour productivity convergence? And to what extent do the results depend on the level of aggregation? To answer these questions we calculated for each sector the unweighted cross-country standard deviation for the log of energy and labour productivity, σ, among the 14 OECD countries (insofar as data are available).[5] Figure 6.1 presents the degree of variation in 'macroeconomic' energy and labour productivity levels, being the sum of aggregate manufacturing, transport, services and agriculture.[6] The figure confirms the preliminary results found in Chapter 5 (Table 5.2) that cross-country differences in energy productivity levels are substantially larger than cross-country differences of labour productivity levels. Moreover it can be seen that over time the standard deviation of the log of energy productivity performance is increasing, indicating σ-divergence, while the opposite is true for cross-country labour productivity performance, displaying a pattern of σ-convergence.

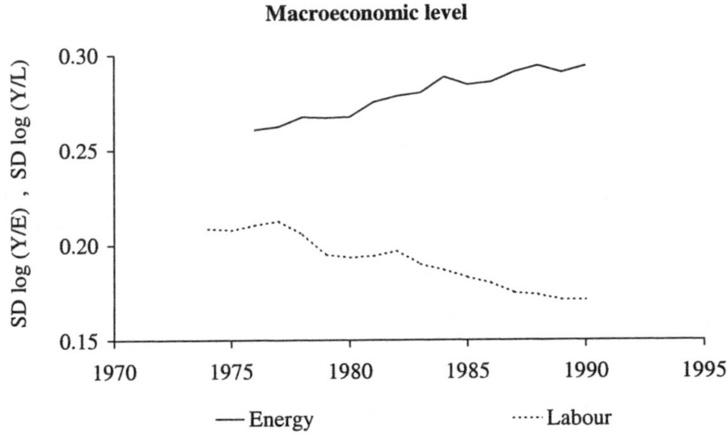

Figure 6.1 Standard deviation of log energy and labour productivity at the macroeconomic level (including aggregate manufacturing, transport, services and agriculture)

In Chapter 5 we found that the level of aggregation matters in examining productivity trends, because of the existence of substantial sectoral heterogeneity in productivity performance. Of course this suggests that convergence patterns may vary among sectors. And indeed in two studies on cross-country productivity convergence across industries, Bernard and Jones (1996a, b) found aggregate outcomes to mask important variation in sectoral productivity movements. Hence we continue by examining the development of cross-country productivity differentials within different sectors.

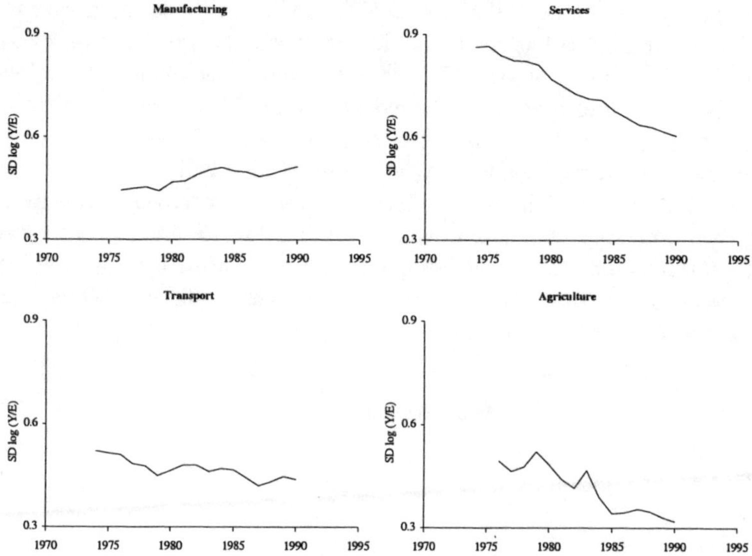

Figure 6.2a Standard deviation of log energy productivity in main sectors

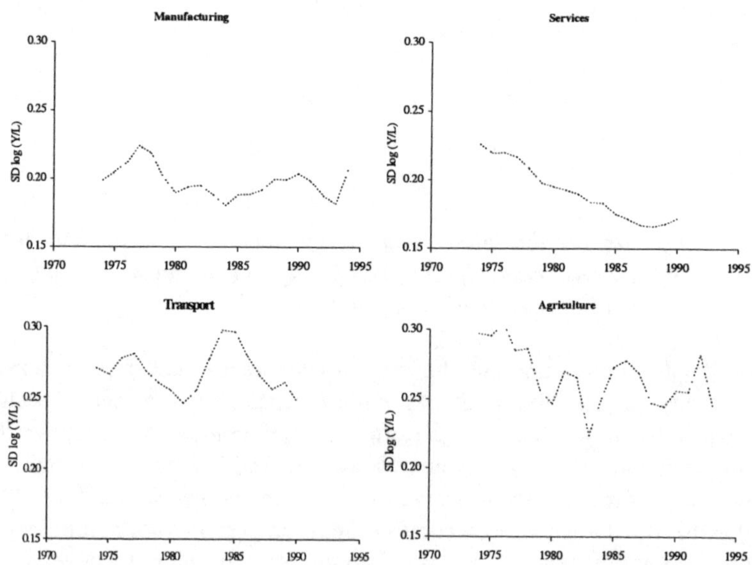

Figure 6.2b Standard deviation of log labour productivity in main sectors

In Figures 6.2a and 6.2b we present the standard deviation of the log of, respectively, energy and labour productivity for aggregate manufacturing, transport, services and agriculture.[7] From Figure 6.2a it can be seen that only manufacturing resembles the macroeconomic pattern of σ-divergence for energy productivity. Transport, agriculture and in particular services display evidence of σ-convergence. Note that the level of cross-country variation is relatively high in services, which is to a large extent due to the exceptional and so far unexplained energy productivity performance of Finland and Italy.[8]

Figure 6.2b shows that the macroeconomic pattern of σ-convergence for labour productivity is only evident in services, and to a lesser extent in the agriculture sector. At the same time, variation in cross-country productivity differentials remains overall fairly constant within aggregate manufacturing and transport (although with fluctuations over time). Comparing Figures 6.2a and 6.2b shows again that in each sector the cross-country variation of energy productivity is substantially larger than that of labour productivity. These results do not change when the United States is removed from the sample. Moreover they accord well with the findings of Bernard and Jones (1996a), who suggest 'that international flows, associated mostly with manufacturing, may not be contributing substantially to convergence either through capital accumulation or technological transfer' (Bernard and Jones 1996a: 1230). Our analysis suggests that this conclusion holds even stronger for energy productivity performance across countries, where international flows cannot prevent an increase in cross-country differences of productivity levels.

This raises the question as to what the determinants of these cross-country productivity differences are. In our search for an answer we subsequently take two steps. First, we go one step further in the σ-convergence analysis than Bernard and Jones (1996a, b) by examining productivity convergence for a breakdown of aggregate manufacturing in order to see whether the energy productivity divergence and the lack of labour productivity convergence observed in aggregate manufacturing is also found within the different manufacturing subsectors. Second, we perform a β-convergence analysis to test whether a statistically significant negative relationship exists between the initial level and the growth rate of productivity performance. Moreover we will try to explain persistent differences in cross-country productivity growth by examining the rôle of different country-specific variables in driving energy and labour productivity growth at the sectoral level. The latter is the subject of section 6.4. Below we continue with a σ-convergence analysis for a breakdown of aggregate manufacturing into ten subsectors.

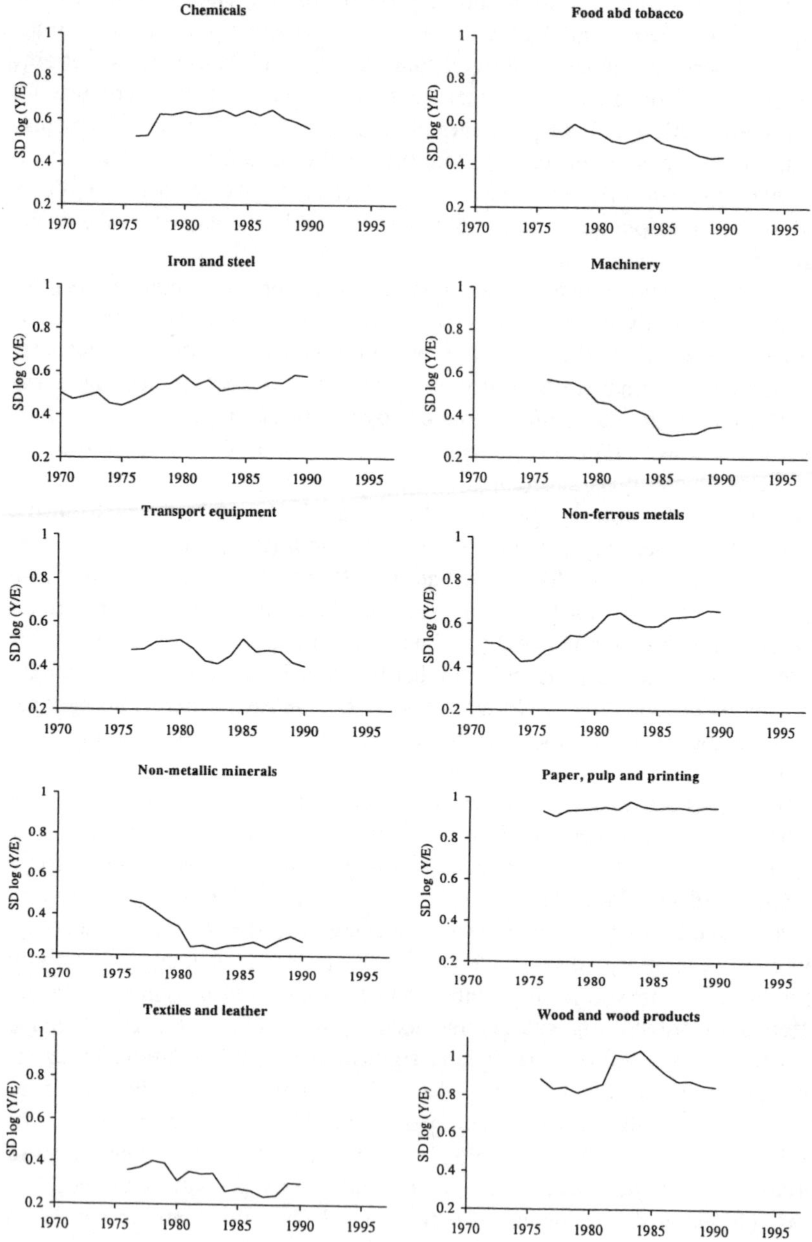

Figure 6.3a Standard deviation of log energy productivity in manufacturing sectors

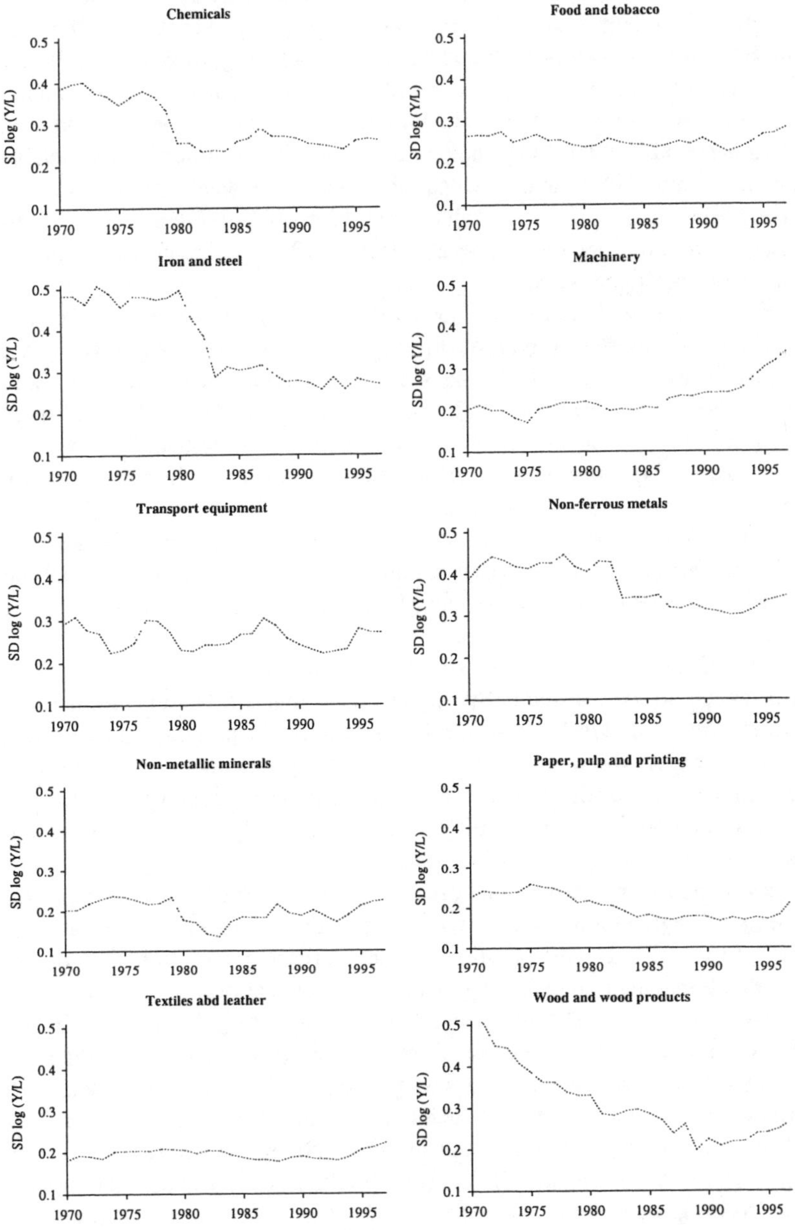

Figure 6.3b Standard deviation of log labour productivity in manufacturing sectors

In Figures 6.3a and 6.3b we present the standard deviation of the log of, respectively, energy and labour productivity for each of the ten manufacturing subsectors included in our dataset.[9] Figure 6.3a reveals that the pattern of divergence in cross-country energy productivity performance at the level of aggregate manufacturing is to be found only in iron and steel and non-ferrous metals. On the contrary, food, machinery, non-metallic minerals (until 1980) and textiles all display evidence of (strong) σ-convergence. Cross-country productivity differences remain more or less constant in non-metallic minerals (after 1980), chemicals, transport equipment, paper and wood.

From Figure 6.3b it can be seen that the lack of labour productivity convergence in aggregate manufacturing is the result of mixed convergence patterns in different manufacturing sectors. Chemicals, iron and steel, non-ferrous metals and wood exhibit (strong) convergence, while machinery shows the opposite pattern of divergence. The food, non-metallic minerals, textile, paper and transport equipment sectors display no clear evidence for either convergence or divergence, although the latter shows substantial fluctuations over time. Moreover, it is to be noted that, in chemicals, iron and steel, non-ferrous metals and non-metallic minerals, convergence of labour productivity performance is particularly strong during the first half of the 1980s.

In conclusion, we found cross-country variation of energy productivity performance to be substantially higher than that of labour productivity performance at all levels of sectoral aggregation, in particular in services, chemicals, paper, wood, and at an ever increasing rate also in iron and steel and non-ferrous metals. In machinery however cross-country variation of energy and labour productivity differences has strongly converged, resulting in a relatively small – although persistent – difference in degree of cross-country variance. Moreover convergence patterns turned out to depend on the level of aggregation, with different sectors displaying varying behaviour: some show reduction in variation, some increasing variation and others neither a clear reduction or increase over the whole period.

These results suggest that different mechanisms may be at work in the different sectors. For example the observed patterns of divergence might be the result of increasing international specialization, while the tendency to converge might be caused by technology spillovers from 'leaders' to 'followers', allowing lagging countries to catch up. Moreover our results suggest that determinants of energy productivity growth and labour productivity growth might differ from each other, since we found no clear-cut (and sometimes even an opposite) relationship between cross-country convergence patterns in terms of energy productivity and labour productivity. Finally, even in those sectors showing evidence of convergence there remain substantial cross-country productivity differences, in particular in terms of energy productivity. This suggests that convergence does not pertain to a uniform steady state for all countries. In order to further examine this issue, we continue in the next section with a search for empirical regularities in the productivity improvements over our cross-section of countries by testing for sectoral patterns of β-convergence. As part of that analysis we will also try to explain differences in energy and labour productivity growth.

6.5 β-CONVERGENCE

This section deals with the notion of convergence in terms of growth rates. In this case energy and labour productivity convergence implies that energy and labour productivity growth rates tend to decline if countries reach their steady states. Obviously, it is not an easy task – if possible at all – to judge whether a country is in its steady state or not. It is possible however to analyse the correlation between growth rates and levels, assuming that a negative correlation between these two provides an indication for convergence, because it suggests that countries with relatively low initial energy and labour productivity levels catch up to more advanced economies. The rationale behind this is that those countries lagging behind in terms of productivity levels can benefit from relatively high marginal factor productivity or from the experience and technologies developed by the countries operating at the forefront. In this section we adopt a panel-data framework to regress average energy and labour productivity growth rates on initial productivity levels, generating an estimate of the coefficient β, for each sector. A negative estimated coefficient β indicates the existence of so-called β-convergence. We start by testing for unconditional convergence, assuming that energy and labour productivity converge towards a unique steady state for all countries included in the data set. Subsequently we relax this assumption and test for conditional convergence, assuming productivity levels to converge towards multiple steady states that are conditional on country-specific characteristics. Finally, as part of this analysis we try to identify the country-specific characteristics that determine (differences in) energy and labour productivity growth across countries. We refer to Appendix 6.B for descriptive statistics of the data used in the subsequent regression analyses.

Table 6.1 *Unconditional β-convergence*

	Energy			Labour		
	β_E	R^2	λ	β_L	R^2	λ
Total	- 0.0368	0.03	0.0075	- 0.1138***	0.30	0.0242
	(0.0246)			(0.0218)		
Agriculture	- 0.3227***	0.33	0.0779	- 0.0575*	0.03	0.0118
	(0.0598)			(0.0300)		
Services	- 0.1432***	0.26	0.0309	- 0.1445***	0.44	0.0312
	(0.0424)			(0.0300)		
Transport	- 0.0827**	0.12	0.0173	- 0.1046**	0.10	0.0221
	(0.0312)			(0.0465)		
Manufacturing	- 0.1524***	0.16	0.0331	- 0.0392	0.00	0.0080
	(0.0459)			(0.0324)		
Chemicals	- 0.0190	0.01	0.0038	- 0.1027***	0.12	0.0217
	(0.0436)			(0.0366)		
Food and Tobacco	- 0.0782**	0.07	0.0163	- 0.1058**	0.12	0.0224
	(0.0385)			(0.0401)		
Iron and Steel	- 0.0442	0.01	0.0090	- 0.1060*	0.05	0.0224
	(0.0557)			(0.0461)		
Machinery	- 0.1729***	0.17	0.0380	- 0.1263**	0.12	0.0270
	(0.0557)			(0.0461)		
Transport Equipment	- 0.3082***	0.21	0.0737	- 0.1480*	0.07	0.0320
	(0.0823)			(0.0834)		
Non-Ferrous Metals	- 0.0153	0.00	0.0031	- 0.0683	0.03	0.0141
	(0.0617)			(0.0494)		
Non-Metallic Minerals	- 0.3156***	0.24	0.0758	- 0.1459***	0.12	0.0315
	(0.0761)			(0.0536)		
Paper, Pulp and Printing	- 0.0435	0.03	0.0089	- 0.0755**	0.06	0.0157
	(0.0355)			(0.0376)		
Textiles and Leather	- 0.3497***	0.19	0.0861	- 0.1577***	0.18	0.0343
	(0.1020)			(0.0433)		
Wood and Wood Products	- 0.0236	0.01	0.0048	- 0.1925***	0.29	0.0428
	(0.0349)			(0.0381)		

Notes:

Standard errors in parentheses.

Asterisks denote levels of significance: *** (1%), ** (5%), * (10%).

Unconditional β-Convergence

We test for unconditional β-convergence by regressing for each sector the growth rate (g) of, respectively, energy and labour productivity (y), on its initial level (and a constant α), generating an estimate of β, according to:

$$g_{it} = \log(y)_{i,t} - \log(y)_{i,t-1} = \alpha + \beta \ln(y)_{i,t-1} + \varepsilon_{it} \qquad (6.1)$$

with i and t denoting, respectively, the cross-country and the time-series dimension, while ε_{it} is the standard error. Following Islam (1995) we use five-year time intervals in order to reduce the influence of business cycle fluctuations and serial correlation of the error term. Hence the growth rate (g) in equation (6.1) is an average over a five-year period (if t = 1975, for example, t–1 = 1970). The results are presented in Table 6.1.

From the table it can be seen that we obtain a statistically significant negative estimate of β for energy productivity growth in most sectors, except for the macroeconomic level (total), chemicals, iron and steel, non-ferrous metals, paper and wood. In terms of labour productivity growth we found β to be statistically significant in all sectors, except for aggregate manufacturing and non-ferrous metals.[10]

These results confirm the findings of Bernard and Jones (1996a) who also report lack of labour productivity convergence in manufacturing, weak evidence for convergence in agriculture and strong evidence in services. It is to be noted however that in most sectors that display evidence of convergence, estimates of β are rather small, indicating that lagging countries catch up only very slowly. Using the estimated values of β, the rate at which the productivity level is converging to a uniform productivity level can be derived (for example Barro and Sala-i-Martin 1992, Mankiw et al. 1992, Islam 1995). Let y^* be the steady-state productivity level and $y(t)$ its actual value at any time t. Approximating around the steady state, the speed of convergence is given by

$$\frac{d \log(y(t))}{dt} = \lambda[\log(y^*) - \log(y(t))] \qquad (6.2)$$

which implies that:

$$\log(y(t)) = (1 - e^{-\lambda t})\log(y^*) + e^{-\lambda t}\log(y(0)) \qquad (6.3)$$

where ($y(0)$) is energy or labour productivity level at some initial date. Subtracting log ($y(0)$) from both sides yields

$$\log(y(t)) - \log(y(0)) = (1 - e^{-\lambda t})[\log(y^*) - \log(y(0))] \quad (6.4)$$

in which $-(1-e^{-\lambda t}) = \beta$. Hence the speed of convergence, λ, is given by $\lambda = -[1/T \log(\beta+1)]$ with T denoting the time interval under consideration.[11] The values of the implied λ are shown in Table 6.2. They confirm the finding of a slow rate of convergence: the time t needed for energy productivity to move halfway its initial level ($y(0)$) and the steady state y^* varies from eight years (textiles) to 225 years (non-ferrous metals); the half life for labour productivity lies in between 16 years (wood) and 87 years (manufacturing).[12]

As previously noted, β-convergence is a necessary but not a sufficient condition for σ-convergence. Our findings confirm that those sectors showing evidence of σ-convergence (see section 6.4) also display evidence of β-convergence: a decreasing cross-country variation of productivity levels implies by definition that countries with relatively low initial energy and labour productivity levels grow relatively fast. However the opposite is not necessarily true, as is illustrated for labour productivity by the machinery, non-metallic minerals and textiles sectors: they pass the test for β-convergence without showing evidence of σ-convergence (see Figure 6.3b).

Finally, it is to be noted that the ability of the simple regression equation (6.1) to explain cross-country productivity growth rates is rather small in most sectors. This is not surprising since the specification of equation (6.1) implicitly builds upon the assumption that energy and labour productivity levels converge towards a uniform steady state. However economies differ and so do (most likely) their steady states. Contrary to a framework of single cross-country regressions, a panel data framework is capable of allowing for cross-country differences in steady states in the form of unobservable individual 'country-effects' (Islam 1995). These country-effects might include all sorts of country-specific tangible and intangible factors that affect productivity growth and which have not been included in equation (6.1) or, to state it differently, have been subsumed in its error term. Therefore in the next section we test for β-convergence allowing for these country-effects.

Conditional β-Convergence

We test for conditional β-convergence by including (unspecified) individual country-effects in equation (6.1), assuming that productivity levels converge to country-specific steady states. Hence, we reformulate equation (6.1) into a panel-data model with individual country effects, according to:

$$g_{it} = \beta \ln(y)_{i,t-1} + \mu_i + \varepsilon_{it} \tag{6.5}$$

with μ_i representing unspecified country-specific (fixed) effects.

In Table 6.2 we present, for each sector, the estimated coefficient β obtained from equation (6.5). The table shows that allowing for individual country-effects substantially improves the explanatory power of the regression equations for both energy- and labour productivity. Moreover in terms of energy productivity, equation (6.5) yields significantly negative estimates of β in all sectors, including now also the macroeconomic level (total), chemicals, iron and steel, non-ferrous metals and machinery. Also in terms of labour productivity the estimates of β are higher in several sectors such as, for example, services and food. The evidence on conditional labour productivity convergence is however less clear-cut than it is for energy productivity convergence: in some sectors, such as for example iron and steel and machinery, allowing for individual country-effects in explaining labour productivity growth yields statistically less significant or even insignificant estimates of β.

This result suggests that in terms of labour productivity the variation in explanatory variables over time is relatively small as compared to cross-country differences, since correcting for the latter by means of including country-specific intercepts results in weaker evidence of a negative relationship between the initial labour productivity level and its growth over time. Nevertheless the regression results suggest that both energy and labour productivity convergence depend to a large extent on individual country-effects, indicating energy and labour productivity to be conditional rather than absolute in virtually all sectors. The latter is illustrated by the fact that the speed of conditional convergence is substantially higher than of unconditional convergence: for energy productivity the half-life, as it follows from the implied λ, now lies between 1 year (transport equipment) and 14 years (total) and for labour productivity it has been reduced to a period in between 47 years (transport equipment) and 77 years (non-ferrous metals).

Of course this brings us back to the question of which are the country-specific variables driving energy and labour productivity growth and hence determining the country-specific steady states. Recall from the introduction that several mechanisms may be at work, causing 'followers' to grow faster than 'leaders': advanced economies may suffer from diminishing returns, lagging countries may benefit from knowledge spillovers, production processes may convergence due to increasing competition, and so on.

Table 6.2 Unspecified conditional β-convergence

	Energy			Labour		
	β_E	R^2	λ	β_L	R^2	λ
Total	- 0.2214*** (0.0691)	0.19	0.0501	- 0.1068*** (0.0262)	0.58	0.0226
Agriculture	- 0.4797*** (0.0888)	0.49	0.1307	- 0.0831* (0.0431)	0.22	0.0174
Services	- 0.2181** (0.1169)	0.44	0.0492	- 0.1783*** (0.0422)	0.80	0.0393
Transport	- 0.6301*** (0.1593)	0.42	0.1989	- 0.1040 (0.1115)	0.16	0.0220
Manufacturing	- 0.6162*** (0.0680)	0.67	0.1915	- 0.0553 (0.0382)	0.16	0.0114
Chemicals	- 0.2620*** (0.0836)	0.33	0.0608	- 0.0929* (0.0484)	0.29	0.0195
Food and Tobacco	- 0.5180*** (0.1292)	0.36	0.1460	- 0.1879*** (0.0542)	0.49	0.0416
Iron and Steel	- 0.3889*** (0.1113)	0.32	0.0985	- 0.0642 (0.1055)	0.11	0.0133
Machinery	- 0.2305* (0.1365)	0.31	0.0524	- 0.0549 (0.0499)	0.45	0.0113
Transport Equipment	- 0.9504*** (0.1127)	0.68	0.6008	- 0.3104** (0.1253)	0.23	0.0743
Non-Ferrous Metals	- 0.5924*** (0.1426)	0.38	0.1795	- 0.0439 (0.0932)	0.10	0.0090
Non-Metallic Minerals	- 0.5087*** (0.0980)	0.55	0.1421	- 0.2089*** (0.0635)	0.39	0.0469
Paper, Pulp and Printing	-0.6513*** (0.1033)	0.60	0.2107	- 0.1053 (0.0783)	0.24	0.0223
Textiles and Leather	- 0.8612*** (0.1285)	0.60	0.3949	- 0.2330*** (0.0545)	0.33	0.0531
Wood and Wood Products	- 1.0637*** (0.1941)	0.60	--	- 0.2298*** (0.0650)	0.47	0.0522

Notes:

Standard errors in parentheses.

Asterisks denote levels of significance: *** (1%), ** (5%), * (10%).

In order to explain persistent differences in cross-country energy and labour productivity growth we replace in equation (6.5) the unspecified country-effects μ_i by a number of country-specific explanatory variables x_i^j, according to:

$$g_{it} = \alpha + \beta \ln(y)_{i,t-1} + \sum_{j=1}^{5} \gamma_j x_{it}^j + \varepsilon_{it} \qquad (6.6)$$

The specified explanatory variables are defined at the sectoral level and include:

Energy prices: $$x_{it}^{1a} = \frac{(p_{E,t} + p_{E,t-1} + p_{E,t-2})}{3}$$

Wages: $$x_{it}^{1b} = \frac{(w_t + w_{t-1} + w_{t-2})}{3}$$

Investment share: $$x_{it}^2 = \frac{I}{Y}$$

Openness: $$x_{it}^3 = \frac{XGS + MGS}{Y}$$

Balassa index: $$x_{it}^4 = \frac{XGS_i \Big/ \sum_{i=1}^{14} XGS_i}{\sum_{s=1}^{10} XGS_{i,s} \Big/ \sum_{i=1}^{14}\sum_{s=1}^{10} XGS_{i,s}}$$

Economies of scale: $$x_{it}^5 = \frac{Y}{\sum_{s=1}^{13} Y}$$

where sectoral indices are omitted for reasons of expositional clarity and with energy prices (x_{it}^{1a}) or wages (x_{it}^{1b}) included, respectively, in the case of explaining energy productivity growth or labour productivity growth. We expect energy prices and wages to be positively correlated with, respectively, energy and labour productivity growth. We took a three year moving average for the energy price and wages to avoid capturing the effect of short-term price fluctuations, assuming that investments in energy- and labour-augmenting technologies do respond to a structural trend in energy price/wage developments rather than to short-term fluctuations. By including the investment share as an explanatory variable we test for the so-called embodiment hypothesis or a vintage effect, assuming that higher investment will contribute to increasing energy and labour productivity growth via technological change embodied in new capital goods (see for example Howarth et al. 1991, Mulder et al. 2003). We expect openness to have a positive impact on productivity growth, since an open sector faces relatively strong competition as well as exchange of knowledge, both of which we assume to have a stimulating effect on productivity growth. The Balassa index is an indicator measuring relative specialization patterns. We expect that if a country specializes in a particular sector, that that sector will be technologically relatively advanced, and hence we expect a positive effect on productivity. Finally, including an indicator for the relative size of a sector within a country captures the potential effect of economies of scale on productivity growth, assuming that a large sector is able to invest relatively much in R&D and in new capital goods, and hence might be a technological leader displaying relatively high productivity growth rates.

The results of regressing average energy productivity growth rates on initial energy productivity levels and these additional explanatory variables, according to equation 6.6, are presented in Table 6.3.[13] It can be seen that the regression analysis generates significantly negative estimates of β in agriculture, transport, iron and steel, machinery, non-metallic minerals, paper and textiles. Compared to Table 6.2 this means that in six sectors the estimate of β is no longer statistically significant once we include the previously mentioned specified explanatory variables. The effect of investment share, openness, specialization and economies of scale on energy productivity growth is mixed and their impact is statistically insignificant in virtually all sectors. An exception is the energy-price effect, which has in all sectors the expected (positive) sign, while the positive impact of energy prices on energy productivity growth is statistically significant in chemicals, iron and steel, and paper, which makes sense since these are energy-intensive sectors. Finally, the speed of convergence has slowed down with the half-life increased to a minimum of five years (wood) and a maximum of 41 years (chemicals).

In Table 6.4 we present the results of regressing average labour productivity growth rates on initial labour productivity levels and the five additional explanatory variables, according to equation (6.6). Table 6.4 shows statistically significantly negative estimates of β for agriculture, food and textiles only. Similar to energy productivity growth, the effect of the investment share, openness, specialization and economies of scale on labour productivity growth is mixed, while their impact is statistically insignificant in virtually all sectors. Or it is statistically significant with an unexpected sign such as, for example, in case of the 'vintage effect' in agriculture, iron and steel, transport equipment and wood, and economies of scale in services.

Finally, the speed of convergence has slowed down with the half life increasing to in between 12 years (textiles) and 95 years (machinery), while for services the estimate yields a positive β, implying divergence.

In sum, while there is strong evidence of conditional convergence in terms of both energy and labour productivity growth (see Table 6.2), we found energy prices and in particular wages, investment share, openness, specialization and economies of scale to play only a limited role in explaining energy and labour productivity growth and hence in determining cross-country productivity differentials. The results shown in Tables 6.3 and 6.4 might however suffer from an omitted variable bias if the included explanatory variables are correlated with the unspecified country-effects μ_i, which are excluded from equation (6.6).

Hence to correct for this potential omitted variable bias, we add the unspecified individual country-effects to equation (6.6), according to

$$g_{it} = \beta \ln(y)_{i,t-1} + \sum_{j=1}^{5} \gamma_j x_{it}^{\ j} + \varepsilon_{it} \tag{6.7}$$

The results of this regression model for energy productivity growth are presented in Table 6.5.

It can be seen that adding the unspecified country effects again affects the estimates substantially. The estimation results in Table 6.5 show that, except for food, non-ferrous metals and wood, all sectors yield a statistically significant estimate of β, with regression equation (6.7) displaying a much better goodness of fit in most sectors than equation (6.6) (see Table 6.3). Except for food and non-ferrous metals, the speed of convergence has increased considerably as compared to Table 6.3, with the half-life to be between one and five years in all other sectors. The statistically significant energy price effect is robust in iron and steel and in paper, while in chemicals the null hypothesis of no effect is only just rejected at the 10 per cent level. In addition energy prices also seem to have a statistically significant positive effect on energy productivity growth in services and textiles. The impact of the other explanatory variables on energy productivity growth remains mixed, with economies of scale having the largest statistically significant effect on energy productivity growth, being positive in services, chemicals and transport equipment, while negative in textiles. We found the vintage effect and openness to have a statistically significant positive effect in transport equipment only, while for specialization this is only the case in iron and steel.

Table 6.3 *Specified conditional β-convergence energy productivity*

	β	Energy Price	Investment share	Openness	Balassa index	Value added share	$\overline{R}^2$	*F*-stat	λ
Agriculture	- 0.1995*	0.0394	- 1.0158			- 2.2995	0.15	1.10	0.0445
	(0.1136)	(0.7225)	(0.8969)			(2.1086)			
Services	- 0.2296	- 0.0147	0.0714			0.3858	0.45	2.25	0.0522
	(0.1791)	(0.5612)	(0.918)			(0.9388)			
Transport	- 0.2350***	0.1168	0.7991**			4.1693**	0.46	3.87	0.0536
	(0.0655)	(0.1439)	(0.3266)			(1.6905)			
Chemicals	- 0.0820	1.3526**	0.9093	- 0.0427	0.1631	- 4.8405	0.28	1.39	0.0171
	(0.0819)	(0.5008)	(1.2271)	(0.0335)	(0.1661)	(8.5819)			
Food	- 0.1211	0.2130	- 0.5277	- 0.0432	0.0113	5.5691	0.35	1.79	0.0258
	(0.0745)	(0.4951)	(0.9092)	(0.0363)	(0.0308)	(3.829)			
Iron and Steel	- 0.3377**	1.9263**	- 0.8636	- 0.0251	0.0068	- 9.5859	0.44	2.79	0.0824
	(0.1252)	(0.7689)	(0.7094)	(0.0184)	(0.1095)	(10.7085)			
Machinery	- 0.2042***	0.4170	0.7337	0.0131	- 0.1112	2.3311	0.43	2.48	0.0457
	(0.0684)	(0.4823)	(1.7063)	(0.0345)	(0.1997)	(1.6439)			
Transport Equipment	- 0.2855	- 0.1642	0.4689	0.0011	0.2225	- 8.192	0.25	0.79	0.0672
	(0.1888)	(0.729)	(1.6648)	(0.0472)	(0.2079)	(13.7238)	0.25	0.79	0.0672
Non-Ferrous Metals	- 0.0844	0.3943	0.0573	- 0.0307	- 0.0943	- 0.4460	0.20	0.75	0.0176
	(0.2229)	(0.5285)	(0.2724)	(0.0208)	(0.1549)	(36.3853)	0.20	0.75	0.0176
Non-Metallic Minerals	- 0.4561***	1.8238	0.0505	- 0.0336	0.2064	- 36.2293*	0.32	1.63	0.1218
	(0.1611)	(1.0916)	(1.1821)	(0.1062)	(0.2016)	(20.1039)	0.32	1.63	0.1218

Table 6.3 *Continued*

	β	Energy Price	Investment share	Openness	Balassa index	Value added share	$\overline{R}^2$	*F*-stat	λ
Paper	- 0.1118*	0.6632**	- 0.8569	- 0.0740	- 0.0643	7.5805	0.47	2.76	0.0237
	(0.0613)	(0.3052)	(0.8666)	(0.0527)	(0.0723)	(6.1778)			
Textiles	- 0.3656*	- 0.4195	0.3092	- 0.0497	0.2136	- 11.7071	0.29	1.29	0.0910
	(0.2053)	(0.783)	(3.1265)	(0.0368)	(0.2621)	(19.7875)			
Wood	- 0.4686	- 0.8418	- 2.7443	- 0.1293*	0.0229	- 53.1902	0.51	1.23	0.1264
	(0.3351)	(0.5508)	(2.2403)	(0.0699)	(0.0667)	(34.2262)			

Notes:

Standard errors in parentheses.
Asterisks denote levels of significance: *** (1%), ** (5%), * (10%).

Table 6.4 Specified conditional β-convergence labour productivity

	β	Wages	Investment share	Openness	Balassa index	Value added share	$\overline{R}^2$	*F*-stat	λ
Agriculture	-0.2090***	0.1649***	-0.4400			-0.2877	0.32	5.16	0.0469
	(0.0522)	(0.0578)	(0.2819)			(0.6738)			
Services	0.0949	-0.0172	0.1203			-0.4534	0.54	5.95	- 0.0181
	(0.0751)	(0.0346)	(0.1489)			(0.2789)			
Transport	-0.6709***	0.3382***	0.1537			4.6484***	0.53	8.47	0.2223
	(0.1248)	(0.0684)	(0.2326)			(1.3658)			
Chemicals	-0.4353***	0.1607***	0.8382	-0.0118	0.0719	-1.2907	0.31	2.62	0.1143
	(0.1394)	(0.0546)							
Food	-0.3867***	0.1509***	-0.9302*	0.0024	-0.0001	1.1356	0.36	4.11	0.0978
	(0.0981)	(0.0474)	(0.4940)	(0.0196)	(0.0165)	(2.3075)			
Iron and Steel	0.1572	-0.3820	-1.2076	-0.0222	-0.2455	-2.7938	0.40	1.98	-0.0292
	(0.9540)	(0.5969)	(2.4004)	(0.0595)	(0.3446)	(118.3932)			
Machinery	-0.6440	0.3208*	-3.6943*	0.0496	-0.2939	-0.0969	0.16	1.65	0.2065
	(0.3952)	(0.1686)	(2.0081)	(0.0304)	(0.1948)	(0.9875)			
Transport Equipment	0.1304	-0.0414	-1.1556	-0.0020	0.3368	-8.2257	0.09	0.83	-0.0245
	(0.3284)	(0.0771)	(0.7465)	(0.0191)	(0.3561)	(7.9319)			
Non-Ferrous Metals	-0.3996**	-0.0328	0.6172	0.0052	-0.0897	142.5101	0.75	5.56	0.1020
	(0.1625)	(0.0654)	(0.5728)	(0.0155)	(0.2068)	(101.2102)			
Non-Metallic Minerals	-0.4768***	0.2121***	-0.7326	0.0771	-0.0064	12.1030	0.17	2.15	0.1296
	(0.1603)	(0.0743)	(0.4954)	(0.0506)	(0.0764)	(7.9561)			

Table 6.4 Continued

	β	Wages	Investment share	Openness	Balassa index	Value added share	$\overline{R}^2$	*F*-stat	λ
Paper	-0.4471***	0.2016**	-0.5374	0.0817**	0.0207	0.6195	0.39	4.50	0.1185
	(0.1564)	(0.0736)	(0.3938)	(0.0316)	(0.0193)	(2.7570)			
Textiles	-0.7421***	0.4614***	-0.0654	-0.0070	-0.0032	-1.1186	0.57	8.21	0.2710
	(0.1434)	(0.1236)	(0.9355)	(0.0114)	(0.0602)	(4.7086)			
Wood	-0.3007	0.1697	0.2561	-0.0491	-0.0233	21.3665***	0.28	2.58	0.0715
	(0.2093)	(0.1369)	(0.5718)	(0.0293)	(0.0177)	(7.6466)			

Notes:

Standard errors in parentheses.
Asterisks denote levels of significance: *** (1%), ** (5%), * (10%).

Table 6.5 Conditional β-convergence energy productivity, specified, fixed effects

	β	Wages	Investment share	Openness	Balassa index	Value added share	$\overline{R}^2$	*F*-stat	λ
Agriculture	- 0.8615***	1.3546	0.2182			12.3662	0.64	9.54	0.3954
	(0.1965)	(1.1801)	(1.1274)			(7.6932)			
Services	- 0.8039***	1.1183*	- 0.5308			7.1791**	0.81	9.88	0.3258
	(0.238)	(0.6375	(0.7828)			(3.2129)			
Transport	- 0.5122*	0.0540	- 0.0171			16.8394	0.80	15.84	0.1436
	(0.2604)	(0.1633)	(0.3651)			(4.3809)			
Chemicals	- 0.7638***	0.7270	1.7622	0.0023	- 0.4818	76.7954***	0.88	19.29	0.2886
	(0.1475)	(0.4416)	(1.0992)	(0.0323)	(0.3812)	(13.9899)			
Food	- 0.0403	1.0346	- 0.5494	- 0.2596**	- 0.5764	- 0.7585	0.63	4.13	0.0082
	(0.3799)	(0.7692)	(1.0185)	(0.1147)	(0.394)	(10.9498)			
Iron and Steel	- 0.8670***	3.8639***	- 1.3162**	0.0052	0.9330***	- 5.4317	0.85	14.58	0.4035
	(0.1679)	(1.2861)	(0.6329)	(0.0310)	(0.2287)	(18.3749)			
Machinery	- 0.5964**	- 0.3718	3.8541	0.1009	0.6565	3.7735	0.66	4.62	0.1815
	(0.2732)	(1.1718)	(3.208)	(0.0971)	(0.8200)	(5.8699)			
Transport Equipment	- 1.1444***	-1.6712***	3.4196**	0.2016***	0.2657	57.3976**	0.84	8.30	---
	(0.2930)	(0.5916)	(1.4195)	(0.0718)	(0.5175)	(26.7348)			
Non-Ferrous Metals	- 0.0262	0.5837	0.0350	- 0.0494	0.0833	132.9891	0.56	2.76	0.0053
	(0.2872)	(1.1114)	(0.429)	(0.0552)	(0.2178)	(80.1969)			
Non-Metallic Minerals	- 0.9527***	0.2167	1.3573	0.0321	- 0.1779	- 17.6544	0.80	10.53	0.6102
	(0.1758)	(1.593)	(1.2800)	(0.1503)	(0.2079)	(32.8798)			

Table 6.5 *Continued*

	β	Energy Price	Investment share	Openness	Balassa index	Value added share	$\overline{R}^2$	F-stat	λ
Paper	- 0.4712**	1.0724*	- 0.9484	- 0.0681	- 0.0111	- 5.335	0.73	5.91	0.1274
	(0.2015)	(0.567)	(1.0232)	(0.1107)	(0.0879)	(19.1876)			
Textiles	- 0.7502***	2.4389*	3.1294	- 0.1738***	0.3466	- 49.3355**	0.79	8.35	0.2774
	(0.2705)	(1.284)	(2.6318)	(0.0544)	(0.2923)	(21.5613)			
Wood	- 0.5676	- 0.9473	- 0.1751	- 0.1839	0.0299	68.6120	0.82	1.76	0.1677
	(1.8578)	(1.054)	(6.9099)	(0.4739)	(0.1620)	(333.4143)			

Notes:

Standard errors in parentheses.
Asterisks denote levels of significance: *** (1%), ** (5%), * (10%).

Table 6.6 *Conditional β-convergence labour productivity, specified, fixed effects*

	β	Wages	Investment share	Openness	Balassa index	Value added share	$\overline{R}^2$	*F*-stat	λ
Agriculture	-0.3109**	0.2547**	-0.4375			0.0781	0.37	2.74	0.0745
	(0.1311)	(0.1241)	(0.4190)			(2.7579)			
Services	-0.0222	-0.2694	0.0056			0.9296	0.70	6.01	0.0045
	(0.1437)	(0.1451)	(0.1603)			(0.8760)			
Transport	-0.9820***	0.4293***	-0.3697*			12.2676***	0.80	11.47	0.8034
	(0.1479)	(0.0928)	(0.2042)			(3.2986)			
Chemicals	-0.6812***	0.1554*	0.9263	0.0025	0.2205	33.3864*	0.59	3.63	0.2286
	(0.1585)	(0.0686)	(0.7407)	(0.0252)	(0.3206)	(18.7647)			
Food	-0.7139***	0.3167***	-1.6230*	0.0076	-0.1620	-3.9158	0.70	6.48	0.2503
	(0.1181)	(0.0472)	(0.4810)	(0.0452)	(0.1493)	(3.4829)			
Iron and Steel	-0.8255**	0.1887	-2.0964	0.0211	1.0264	-17.7137	-0.03	0.96	0.3491
	(0.3652)	(0.2047)	(2.1893)	(0.0582)	(0.6392)	(50.2751)			
Machinery	-0.5521**	0.3270***	-0.1938	-0.0362	0.5117	-0.5526	0.85	11.25	0.1606
	(0.2132)	(0.0887)	(1.1622)	(0.0302)	(0.3691)	(3.2448)			
Transport Equipment	0.0673	-0.0441	-1.6555	0.0022	1.4651	-13.3608	-0.12	0.86	-0.0130
	(0.5017)	(0.2107)	(1.1400)	(0.0408)	(0.9513)	(14.5390)			
Non-Ferrous Metals	-0.1825	0.0517	-0.6133	-0.0274	0.0701	86.3529	0.20	1.37	0.0403
	(0.3780)	(0.0961)	(0.8285)	(0.0336)	(0.3289)	(104.5311)			
Non-Metallic Minerals	-1.0523***	0.3783***	-0.6395	0.2196**	0.0242	3.4617	0.50	3.35	--
	(0.1999)	(0.0786)	(0.4777)	(0.0814)	(0.2161)	(13.5373)			

Table 6.6 Continued

	β	Wages	Investment share	Openness	Balassa index	Value added share	$\overline{R}^2$	*F*-stat	λ
Paper	-0.6466***	0.2485***	0.2950	0.0743**	-0.0155	18.5136	0.78	9.60	0.2080
	(0.1471)	(0.0602)	(0.2997)	(0.0328)	(0.0289)	(4.9143)*			
Textiles	-1.0944***	0.7292***	0.4599	0.0126	-0.1376	3.4939	0.72	6.097	--
	(0.1727)	(0.1509)	(0.8576)	(0.0179)	(0.0899)	(7.4314)			
Wood	-0.9013***	0.3665***	-0.1888	0.0004	-0.0520**	20.9609	0.76	7.32	0.4632
	(0.2419)	(0.1107)	(0.5976)	(0.0292)	(0.0193)	(17.9317)			

Notes:

Standard errors in parentheses.
Asterisks denote levels of significance: *** (1%), ** (5%), * (10%).

In Table 6.6 we present the regression results of equation (6.7) for labour productivity growth. It can be seen that in most sectors labour productivity growth is also better explained if we account for country-specific fixed effects. Moreover the speed of convergence increased in most sectors as compared to Table 6.5, with the half-life between four years (transport) and 36 years (agriculture), and with machinery as the most important exception (with its half life increasing to 367 years). Table 6.7 shows that wages have a statistically significant positive effect on labour productivity growth in agriculture and textiles, while this is negative in transport equipment. Moreover, similar to energy productivity growth, the impact of the other explanatory variables on labour productivity growth remains mixed, with economies of scale again having the largest statistically significant effect, being positive in transport, non-ferrous metals and paper, while it is again negative in textiles. We found openness to have a statistically significant positive effect in paper only, while for specialization this is again only the case in iron and steel. Finally, the results do not give any support to the vintage effect, with the only statistically significant estimates displaying a negative sign.

6.6 CONCLUSIONS

The introduction of new or endogenous growth theories generated renewed interest in the question of whether income and/or productivity levels across countries tend to converge, for example due to capital accumulation or technology transfers. In this chapter we added to the existing empirical analyses of convergence patterns a systematic comparison of energy and labour productivity convergence at a detailed sectoral level for 14 OECD countries, covering the period 1970–97. A σ-convergence analysis revealed that the development of the cross-country variation in energy and labour productivity performance depends on the level of aggregation, with different patterns of productivity convergence and divergence across sectors. At the macroeconomic level we found evidence of energy productivity divergence, driven by aggregate manufacturing, as well as labour productivity convergence, mainly driven by services. The manufacturing energy productivity divergence turns out to be caused by the iron and steel and the non-ferrous metals sectors. Moreover despite a lack of evidence of labour productivity convergence at the aggregate manufacturing level, there is evidence of labour productivity convergence in several manufacturing subsectors, with machinery as the most important exception in that it shows a clear pattern of divergence (in particular after 1985). A β-convergence analysis, using a panel-data approach, led to the conclusion that in most sectors energy productivity growth is relatively high in countries with relatively low initial productivity levels, while in several sectors this is also true for labour productivity. This result supports the hypothesis that relatively backward countries tend to catch up to more advanced economies, in particular in terms of energy productivity, possibly because they can benefit from the experience and technologies developed by the countries operating at the forefront.

However in spite of the evidence of convergence, cross-country differences in energy and labour productivity performance seem to be persistent. Our β-convergence analysis has shown convergence to be conditional on cross-country differences in steady-state characteristics, rather than to be unconditional with productivity levels converging to a uniform steady state for all countries. This is in line with the results of our σ-convergence, which indicated that cross-country differences in productivity levels remain to exist, even in those sectors that display a convergence pattern. Moreover we found that the speed of energy productivity convergence is in general higher than the speed of labour productivity convergence, in particular if we account for (unspecified) country-specific characteristics. Nevertheless at the same time cross-country differences in energy productivity levels were found to be still substantially larger than cross-country differences in labour productivity levels at all levels of sectoral aggregation.

In our search for the country- and sector-specific fundamentals determining these (differences in) energy and labour productivity developments, we found energy prices to stimulate energy productivity growth in the energy intensive sectors while we did not find much evidence of a positive relationship between wages and labour productivity growth. Moreover our data show the cross-country differences in wages to be considerably larger than cross-country differences in final energy prices (measured by the standard deviation of the log of each variable). Hence they are not likely to explain the persistent relatively high cross-country differences in energy productivity levels as compared to labour productivity levels. In addition we found economies of scale to contribute to energy and labour productivity growth in several sectors, while investment share, openness and specialization play only a very limited role in explaining cross-country differences in energy and labour productivity growth. Hence there is a need for additional research to further explain sectoral trends in energy and labour productivity growth across countries.

A possible explanation for the relatively high variation in energy productivity levels across countries might be that cross-country differences in environmental awareness (influenced by social pressure) or stringency of environmental policies cause energy-efficiency improvements to be a matter of urgency to different degrees in different countries. Another reason might be a lack of international diffusion of energy-saving technologies as compared to technologies enhancing labour productivity. This can be caused by the fact that, in contrast with labour costs, in most sectors energy costs form only a small part of total production costs, and hence firms do not have the incentive to search for best-practice technologies in the international market, as opposed to labour-augmenting technologies.

In any case, since productivity growth is primarily driven by technological change, our results suggest that patterns of international technology flows do exist, while at the same time they seem to be limited and at least to some extent sector-specific. There is some reason to believe that technology diffusion and knowledge spillovers are local rather than global (cf. Keller 2002). This suggests the need for specific attention to be paid to the spatial dimension of technology diffusion in driving energy and labour productivity growth in different countries.

In conclusion, the empirical analyses presented in this part of the book make clear that a careful examination of changes in energy efficiency demands a sufficient degree of sectoral detail. Therefore in the applied economic policy analysis presented in the final part of this volume, we do allow for a high degree of sectoral diversity when we assess the impact of technology diffusion on aggregate energy demand.

APPENDIX 6.A

Table 6.A.1 Unconditional β-convergence, period dummies

	Energy			Labour		
	β_E	R^2	λ	β_L	R^2	λ
Total	-0.0193	0.26	0.0039	-0.1127***	0.28	0.0239
	(0.0223)			(0.0233)		
Agriculture	-0.3109***	0.38	0.0745	-0.0636*	0.07	0.0131
	(0.0585)			(0.0372)		
Services	-0.1427***	0.35	0.0308	-0.1209***	0.53	0.0258
	(0.0445)			(0.0331)		
Transport	-0.0832***	0.25	0.0174	-0.1665***	0.39	0.0364
	(0.0301)			(0.0462)		
Manufacturing	-0.1500***	0.12	0.0325	-0.0311	0.01	0.0063
	(0.0474)			(0.0350)		
Chemicals	-0.0172	0.10	0.0035	-0.1678***	0.33	0.0367
	(0.0452)			(0.0484)		
Food and Tobacco	-0.0804**	0.18	0.0168	-0.0755	0.19	0.0157
	(0.0377)			(0.0486)		
Iron and Steel	-0.0489	0.08	0.0100	-0.1965***	0.40	0.0438
	(0.0520)			(0.0540)		
Machinery	-0.1969***	0.31	0.0439	-0.2172***	0.25	0.0490
	(0.0544)			(0.0623)		
Transport Equipment	-0.2833***	0.31	0.0666	-0.1534	0.26	0.0333
	(0.0808)			(0.0951)		
Non-Ferrous Metals	-0.0243	0.30	0.0049	-0.1369**	0.25	0.0294
	(0.0551)			(0.0517)		
Non-Metallic Minerals	-0.3002***	0.33	0.0714	-0.1639**	0.15	0.0358
	(0.0764)			(0.0777)		
Paper, Pulp and Printing	-0.0417	0.21	0.0085	-0.0796*	0.22	0.0166
	(0.0336)			(0.0401)		
Textiles and Leather	-0.3204***	0.33	0.0773	-0.1351**	0.23	0.0290
	(0.0998)			(0.0731)		
Wood and Wood	-0.0180	0.19	0.0036	-0.2028***	0.32	0.0453
	(0.0343)			(0.0461)		

Notes:

Standard errors in parentheses.

Asterisks denote levels of significance: *** (1%), ** (5%), * (10%).

APPENDIX 6.B

Table 6.B.1–6.B.14 *Descriptive sectoral statistics*

Total	ln(Y/E)	ln(Y/L)	ln(Y)	Manufacturing	ln(Y/E)	ln(Y/L)	ln(Y)
Mean	1.50	10.39	26.22	Mean	0.99	10.35	24.97
Median	1.55	10.39	26.29	Median	0.94	10.33	24.70
Maximum	2.14	10.91	29.31	Maximum	1.89	11.24	27.98
Minimum	0.66	9.65	23.96	Minimum	-0.85	9.70	22.78
Std. Dev.	0.34	0.26	1.39	Std. Dev.	0.52	0.31	1.41
Skewness	-0.22	-0.28	0.18	Skewness	-0.30	0.11	0.20
Kurtosis	2.37	2.66	2.10	Kurtosis	2.72	2.41	1.89
Observations	300	339	355	Observations	328	385	386

Agriculture	ln(Y/E)	ln(Y/L)	ln(Y)	Wage	P_{energy}	I/Y	Y_i/Y
Mean	15.66	9.70	23.04	0.04	0.23	0.25	0.04
Median	15.68	9.73	23.05	0.02	0.22	0.24	0.04
Maximum	18.45	10.74	25.58	0.39	0.45	0.52	0.14
Minimum	14.20	8.44	21.24	0.00	0.08	0.08	0.02
Std. Dev.	0.52	0.46	1.14	0.06	0.08	0.07	0.02
Skewness	0.43	-0.22	0.29	2.82	0.38	0.66	1.66
Kurtosis	6.12	2.73	1.97	13.11	2.69	3.56	5.85
Observations	344	370	382	366	279	349	217

Services	ln(Y/E)	ln(Y/L)	ln(Y)	Wage	P_{energy}	I/Y	Y_i/Y
Mean	16.80	10.54	25.71	2.34	0.47	0.29	0.52
Median	16.79	10.52	25.93	0.60	0.39	0.28	0.52
Maximum	18.06	10.91	28.93	18.65	1.22	0.52	0.68
Minimum	15.51	9.88	23.08	0.05	0.19	0.14	0.36
Std. Dev.	0.60	0.22	1.46	4.49	0.23	0.08	0.08
Skewness	0.14	-0.43	0.23	2.28	1.21	0.48	-0.01
Kurtosis	2.27	3.03	2.48	6.86	3.60	2.85	2.33
Observations	214	184	245	179	279	155	217

Transport	ln(Y/E)	ln(Y/L)	ln(Y)	Wage	P_{energy}	I/Y	Y_i/Y
Mean	13.54	10.41	23.47	0.20	0.22	0.29	0.06
Median	13.66	10.42	23.56	0.08	0.20	0.29	0.05
Maximum	14.41	11.13	26.13	1.48	0.92	0.54	0.11
Minimum	12.56	9.39	21.52	0.01	0.00	0.13	0.04
Std. Dev.	0.43	0.32	1.14	0.31	0.11	0.08	0.02
Skewness	-0.65	-0.36	0.28	2.54	2.99	0.17	0.74
Kurtosis	2.44	3.18	2.48	8.58	17.37	2.41	3.03
Observations	297	262	300	257	281	239	217

Chemicals	ln(*Y/E*)	ln(*Y/L*)	ln(*Y*)	Wage	P_{ENERGY}	*I/Y*	Open	Balassa	Y_i/Y
Mean	13.65	10.73	22.52	0.08	0.28	0.24	2.45	0.94	0.02
Median	13.71	10.75	22.53	0.03	0.29	0.20	2.02	0.94	0.02
Maximum	15.68	11.75	25.60	0.57	0.56	0.94	9.58	1.91	0.13
Minimum	12.26	9.17	19.57	0.00	0.00	0.11	0.10	0.29	0.01
Std. Dev.	0.65	0.45	1.55	0.12	0.10	0.12	1.82	0.37	0.01
Skewness	0.39	-0.41	-0.01	2.47	-0.26	2.40	0.93	0.43	5.72
Kurtosis	3.54	3.43	1.96	9.00	3.87	10.48	3.45	2.55	59.28
Observations	347	343	364	344	287	298	345	351	218

Food	ln(*Y/E*)	ln(*Y/L*)	ln(*Y*)	Wage	P_{ENERGY}	*I/Y*	Open	Balassa	Y_i/Y
Mean	15.66	10.53	23.04	0.11	0.37	0.15	1.15	1.28	0.04
Median	15.72	10.56	23.16	0.06	0.34	0.15	0.71	0.97	0.04
Maximum	16.69	11.20	25.54	0.59	0.99	0.30	5.48	5.10	0.14
Minimum	14.55	9.56	20.99	0.01	0.20	0.08	0.04	0.06	0.02
Std. Dev.	0.50	0.30	1.34	0.14	0.13	0.05	1.16	1.14	0.01
Skewness	0.09	-0.27	0.08	2.08	2.27	0.77	1.79	1.64	4.43
Kurtosis	2.09	3.11	1.76	6.69	9.20	3.27	5.65	5.18	45.47
Observations	328	337	364	332	270	310	351	351	218

Iron and Steel	ln(*Y/E*)	ln(*Y/L*)	ln(*Y*)	Wage	P_{ENERGY}	*I/Y*	Open	Balassa	Y_i/Y
Mean	13.49	10.39	21.65	0.05	0.29	0.24	3.01	0.98	0.01
Median	13.59	10.48	21.59	0.02	0.27	0.20	1.83	0.89	0.01
Maximum	14.79	11.46	24.51	0.34	0.75	1.54	16.71	3.02	0.03
Minimum	11.99	9.18	17.96	0.00	0.12	0.00	0.09	0.14	0.00
Std. Dev.	0.56	0.49	1.77	0.07	0.12	0.17	3.34	0.58	0.01
Skewness	-0.61	-0.30	-0.19	1.90	1.23	3.28	2.04	1.01	1.29
Kurtosis	2.90	2.41	2.05	6.09	4.46	20.60	7.19	3.76	5.00
Observations	353	343	364	344	279	322	351	351	218

Machinery	ln(*Y/E*)	ln(*Y/L*)	ln(*Y*)	Wage	P_{ENERGY}	*I/Y*	Open	Balassa	Y_i/Y
Mean	16.62	10.26	23.63	0.32	0.41	0.11	1.87	0.88	0.08
Median	16.64	10.28	23.37	0.06	0.37	0.11	1.44	0.91	0.07
Maximum	17.99	11.19	27.01	2.37	1.36	0.20	8.24	1.61	0.27
Minimum	15.64	8.98	20.95	0.01	0.18	0.05	0.17	0.38	0.04
Std. Dev.	0.50	0.36	1.49	0.48	0.19	0.03	1.51	0.28	0.03
Skewness	0.09	-0.63	0.15	1.99	2.75	0.38	1.40	0.01	2.32
Kurtosis	2.36	3.80	1.94	6.49	11.82	2.81	4.90	2.08	13.10
Observations	308	315	362	304	270	305	350	351	218

Transport eq.	ln(*Y/E*)	ln(*Y/L*)	ln(*Y*)	Wage	P_{ENERGY}	*I/Y*	Open	Balassa	Y_i/Y
Mean	16.72	10.44	22.63	0.17	0.47	0.15	3.19	0.87	0.03
Median	16.71	10.43	22.84	0.08	0.38	0.15	2.25	0.89	0.03
Maximum	17.89	11.05	25.57	0.98	1.88	0.40	16.27	1.85	0.12
Minimum	15.72	9.78	19.90	0.00	0.09	0.05	0.19	0.22	0.01
Std. Dev.	0.48	0.27	1.58	0.24	0.33	0.05	2.85	0.38	0.01
Skewness	0.20	0.18	0.02	2.05	2.11	0.83	1.59	0.41	1.56
Kurtosis	2.26	2.24	1.78	6.26	7.21	4.93	5.63	2.62	13.03
Observations	338	240	364	229	270	222	351	351	218

Non-ferr. met.	ln(*Y/E*)	ln(*Y/L*)	ln(*Y*)	Wage	P_{ENERGY}	*I/Y*	Open	Balassa	Y_i/Y
Mean	14.03	10.52	20.82	0.02	0.50	0.22	4.04	1.36	0.00
Median	14.11	10.54	21.03	0.01	0.47	0.17	3.02	0.85	0.00
Maximum	15.17	11.58	23.66	0.13	1.08	1.04	20.18	8.68	0.01
Minimum	12.50	8.93	16.87	0.00	0.13	0.06	0.16	0.11	0.00
Std. Dev.	0.61	0.53	1.65	0.03	0.19	0.14	3.83	1.55	0.00
Skewness	-0.70	-0.46	-0.34	2.47	0.62	1.88	1.98	2.60	0.15
Kurtosis	3.00	2.77	2.69	8.29	3.09	8.08	7.21	10.05	2.97
Observations	321	343	364	344	272	322	351	351	218

Non-met. min.	ln(*Y/E*)	ln(*Y/L*)	ln(*Y*)	Wage	P_{ENERGY}	*I/Y*	Open	Balassa	Y_i/Y
Mean	14.23	10.43	21.86	0.05	0.22	0.18	0.74	0.91	0.01
Median	14.24	10.44	21.63	0.02	0.21	0.18	0.50	0.71	0.01
Maximum	15.25	11.11	24.11	0.20	0.42	0.37	3.18	4.04	0.03
Minimum	13.18	9.74	19.40	0.00	0.08	0.07	0.03	0.18	0.00
Std. Dev.	0.34	0.28	1.39	0.05	0.07	0.06	0.65	0.66	0.00
Skewness	-0.29	-0.03	-0.02	1.36	0.53	0.61	1.38	2.04	0.73
Kurtosis	3.08	2.40	1.56	4.08	3.11	3.31	4.41	8.18	4.40
Observations	340	315	364	332	279	282	351	351	218

Paper	ln(*Y/E*)	ln(*Y/L*)	ln(*Y*)	Wage	P_{ENERGY}	*I/Y*	Open	Balassa	Y_i/Y
Mean	15.11	10.52	22.73	0.11	0.44	0.17	0.89	1.67	0.04
Median	15.44	10.46	22.96	0.05	0.41	0.15	0.66	0.72	0.03
Maximum	16.51	11.73	25.53	0.80	1.15	0.49	4.18	8.59	0.11
Minimum	13.35	9.82	20.98	0.01	0.22	0.07	0.02	0.11	0.02
Std. Dev.	0.89	0.38	1.26	0.19	0.16	0.07	0.78	1.97	0.01
Skewness	-0.79	1.07	0.41	2.90	1.35	1.37	1.32	1.97	1.22
Kurtosis	2.12	4.24	2.40	10.00	5.49	5.07	4.68	6.30	5.32
Observations	314	337	363	332	279	285	351	351	218

Textiles	ln(*Y/E*)	ln(*Y/L*)	ln(*Y*)	Wage	P_{ENERGY}	*I/Y*	Open	Balassa	Y_i/Y
Mean	15.99	9.86	22.14	0.08	0.37	0.10	2.88	0.98	0.02
Median	15.99	9.88	22.13	0.03	0.35	0.09	1.67	0.91	0.02
Maximum	16.83	10.46	24.78	0.43	0.74	0.23	14.05	3.71	0.05
Minimum	15.29	9.09	18.98	0.00	0.16	0.00	0.12	0.14	0.00
Std. Dev.	0.33	0.31	1.66	0.10	0.10	0.03	3.15	0.73	0.01
Skewness	0.14	-0.29	-0.14	1.85	0.95	0.99	1.71	1.69	1.01
Kurtosis	2.64	2.57	1.68	6.12	3.85	4.40	5.24	6.18	5.04
Observations	319	337	364	332	279	310	351	351	218

Wood	ln(*Y/E*)	ln(*Y/L*)	ln(*Y*)	Wage	P_{ENERGY}	*I/Y*	Open	Balassa	Y_i/Y
Mean	15.71	10.01	21.30	0.03	0.44	0.15	1.49	1.62	0.01
Median	15.60	10.06	21.44	0.01	0.40	0.14	0.99	0.58	0.01
Maximum	17.51	10.65	24.25	0.20	1.18	0.60	9.37	10.51	0.03
Minimum	14.11	8.30	18.37	0.00	0.08	0.04	0.05	0.01	0.00
Std. Dev.	0.87	0.38	1.30	0.04	0.20	0.07	1.62	2.25	0.01
Skewness	0.12	-1.24	0.11	2.63	0.95	1.76	2.44	1.92	1.09
Kurtosis	1.96	5.47	2.52	9.05	4.27	11.06	9.71	5.93	2.80
Observations	254	346	364	323	260	283	351	351	218

NOTES

1. Although Dollar and Wolff (1988, 1993) distinguish 28 sectors, they only present a labour productivity convergence indicator for a few years and did not perform a regression analysis to test for convergence patterns.
2. For more complete surveys of the convergence debate we refer to Barro and Sala-i-Martin (1995), Broadberry (1996), Durlauf and Quah (1999), Fagerberg (1994), *Economic Journal* (1996) and Islam (2003). For more recent work on evidence of and driving forces behind convergence patterns, see for example Baumol et al. (1994), van Ark and Crafts (1996), Kumar and Rusell (2002), Miller and Upadhyah (2002) and Tondl (2001) among many others.
3. See Bernard and Durlauf (1996) and Durlauf and Quah (1999) for further discussion of empirical methodological issues of convergence tests.
4. See Chapter 5, section 5.2 for a brief discussion on the appropriateness of using PPPs. For a discussion of this issue in empirical analysis of convergence at the sectoral level see Sørensen (2001) and Bernard and Jones (2001).
5. In the literature on convergence analysis, two measures for σ-convergence are used interchangeably: (1) the SD log of per capita income or productivity (y), and (2) the coefficient of variation which equals the SD of per capita income or productivity divided by the sample average. They are defined, respectively, as:

(1) $\sqrt{\frac{1}{n}\sum_{i=1}^{n}\left(\log y_i - \overline{\log y}\right)^2}$, $\overline{\log y} = \frac{1}{n}\sum_{i=1}^{n}\log y_i$

(2) $\sqrt{\frac{1}{n}\sum_{i=1}^{n}\left(\frac{y_i - \bar{y}}{\bar{y}}\right)^2}$, $\bar{y} = \frac{1}{n}\sum_{i=1}^{n} y$.

Dalgaard and Vastrup (2001) show that these measures lead to different conclusions when applied to the data set from the Penn World Table used in Jones (1997), caused by the fact that the measures assign different weights to individual countries' performance. We have therefore used both measures in our convergence analysis, finding both measures to yield an identical pattern of convergence, although with small differences in the size of cross-country variance. Details are available upon request. Here, we only present the result of the SD log-measure (1).

6. Due to limited data availability the calculation of cross-country dispersion, as shown in Figure 6.1, excludes Canada, Japan, the Netherlands and Sweden.
7. Due to limited data availability, the following countries are not included in the calculation of cross-country dispersion, shown in Figure 6.2. Manufacturing: Japan, the Netherlands; agriculture: Japan, the Netherlands; services: the Netherlands, Sweden; transport: Canada, the Netherlands.
8. Excluding Finland and Italy from the sample for services reduces the cross-country dispersion by about 40 per cent while leaving the pattern of σ-convergence unchanged. Note that the Netherlands also exhibits an exceptional development of energy productivity performance in services, but has already been excluded form the sample used in Figure 6.2a. For further details see Chapter 5.
9. Due to limited data availability, the following countries are excluded from the calculation of cross-country dispersion, shown in Figure 6.3. Food: Australia and Canada; iron and steel: Japan; machinery: Canada, Japan, the Netherlands; transport equipment: Canada; non-ferrous metals: Denmark; paper: Australia, Japan; textiles: Canada; wood: Canada, France, Japan, the United Kingdom, the United States.
10. We also estimated equation (6.1) including a period-specific fixed effect ηt according to $git = \alpha + \beta \ln(y)i,t{-}1 + \eta t + \varepsilon it$. The regression results with these period dummies included do not substantially improve the estimates in most sectors, except for non-ferrous metals and in terms of labour productivity also for chemicals, iron and steel and machinery. These findings suggest that in spite of a few exceptions, in general there is not much evidence for substantial differences in growth rates between the time periods included. Details are available upon request.
11. Since we use five-year time intervals, $T = 5$ in our analysis. Note that in Islam (1995) $\lambda = -[(1/T)\ln(\beta)]$ due to the fact that he takes $\ln(y)it$ instead of $[\ln(y)it - \ln(y)it\text{-}1]$ as the dependent variable.
12. The half-life (H) is derived from $e{-}\lambda H = 0.5 \Leftrightarrow H = \ln(2) / \lambda$
13. We also controlled for different specifications of energy prices (current prices, five-year moving average, and log three-year and log five-year moving average), investment share $((I/Y)t{-}1$, (I/K), $(I/K)t{-}1$ and $\ln(I/K)t{-}1)$, as well as an interaction term of investment share and log initial energy productivity $(\ln(Y/E)0{*}\ (I/Y))$. All these specifications did not substantially alter the estimates. Details are available upon request.

PART IV

Policy Analyses and Conclusions

7. Dynamics of technology diffusion in an applied energy–economy model for the Netherlands

7.1 INTRODUCTION

Increasing energy efficiency has been a concern for policy-makers ever since the energy crisis of the 1970s. Nowadays concern over energy efficiency is mainly due to the environmental externalities resulting from increased energy use. Technological change plays a crucial role in improving energy efficiency, and hence understanding the interaction between technological change, energy use and environmental policy is a prerequisite for designing effective policies as well as to properly forecast their impact. So far the debate over the role of (endogenous) technological change in economic models of energy–economy–climate interaction mainly centers around the issues of innovation and learning-by-doing at a very high level of aggregation (see, for example, Goulder and Schneider 1999, Goulder and Mathai 2000, Nordhaus 2002). It is obvious however that the interaction between environmental policy and choice of technology depends ultimately on the dynamic aspects of firms' decisions regarding the use of different technologies. As stated in Chapter 4, this implies that 'the challenge for governments is to develop policies aimed at achieving the macroeconomic or generic goals imposed by emission reduction targets, while taking into account the microeconomic characteristics that determine adoption behaviour of individual firms'. Since policy-makers often use applied economic models in developing such policies, these models should ideally include an appropriate representation of microeconomic adoption and diffusion processes. An incomplete specification of processes of technological change is likely to produce biased estimates of energy demand and of the impact of policies aiming to increase energy efficiency.

In this chapter we make an effort in this direction by examining the effect of several microeconomic characteristics of technology adoption and diffusion on sectoral energy consumption. For this purpose we use an existing applied economic model of economy–energy interaction, used for policy analysis in the Netherlands. The model is characterized by a vintage structure and allows for a high degree of sectoral detail. The adoption and diffusion mechanisms in the policy model are evaluated, reinterpreted and modified, building upon insights obtained by the theoretical models developed in Part II of this volume (Chapters 3 and 4). In sum, we introduce into the model mechanisms representing technological complementarity as well as a learning-by-using process for new technologies, and we refine the (external) learning or diffusion process by which technologies are upgraded. Finally, we examine the impact of environmental policies on energy consumption, under different assumptions on adoption and diffusion processes, given sector-specific characteristics of technological change.[1]

The research agenda for this chapter follows from the conclusions drawn in the previous chapters. Throughout this volume we have analysed the aggregate effects of microeconomic characteristics of adoption and diffusion processes, such as for example the role of uncertainty in the presence of irreversible investment, the embodiment of technological change in capital goods, learning-by-using processes and complementarities between technologies. It has been argued that these characteristics are among the important determinants of adoption behaviour and as such can help to explain the observed slow diffusion of energy-saving technologies. At the same time, it has been shown that neglecting some of these issues tends to result in a biased estimation of the effectiveness of environmental policies, such as energy taxes and investment subsidies (see Chapters 3 and 4). The empirical part of this study showed that patterns of energy consumption differ considerably among sectors (see Chapters 5 and 6). The empirical picture thereby supports the idea that a careful examination of changes in aggregate energy use asks for a sufficient degree of sectoral detail. Moreover it suggests that changes in the aggregate energy–output ratio are ultimately driven by sector-specific patterns of technological change, rather than by an increasing uniform stock of knowledge or an exogenous and constant Autonomous Energy Efficiency Improvement (AEEI).

The aim of this chapter is to investigate the possible relevance of these insights in the context of an existing applied economic model. The models developed in Chapters 3 and 4 were characterized by a rather detailed, microeconomic characterisation of adoption behaviour building upon recent theoretical insights, but also by a lack of empirical validation, as a result of which they rely on numerically chosen parameter values only. In contrast, applied economic policy models are often characterized by a more stylized and aggregate representation of technological change, building upon traditional theoretical insights, but relying on empirically estimated parameter values. Hence an important question arises as to what extent the results of applied economic policy models are sensitive to theoretically elegant mechanisms such as are developed in stylized theoretical models.

To examine these issues we use the Netherlands Energy Demand Model, NEMO, developed and used by the CPB Netherlands Bureau for Economic Policy Analysis (Koopmans et al. 1999). This applied economic model has been developed to link energy use to energy prices and economic growth in order to analyse energy-saving and greenhouse gas emission reduction policies. NEMO describes the substitution between capital, labour and energy in the production of energy services, assuming exogenous technological progress, but emphasizing the development of energy efficiency due to adoption and diffusion of energy-saving technologies. With respect to the latter, it is to be noted that NEMO aims to bridge conflicting predictions on the relationship between economic growth and energy use that result from bottom-up and top-down models (Koopmans and te Velde 2001). Compared to macroeconomic top-down models, bottom-up models often suggest a lower future energy–output ratio by emphasizing the existence of significant unexploited opportunities for cost-effective investments in energy-saving (de Beer 1998, IWG 1997). For this reason, NEMO deals explicitly with the energy efficiency gap – as do the theoretical models developed in Chapters 3 and 4. It does so mainly by using a bottom-up engineering database to estimate the key parameters of the model, defining the rate of technological progress, substitution possibilities and price elasticities at a sectoral level.

This chapter is organized as follows. First, section 7.2 provides a brief description of NEMO and the way its key parameters are calibrated. Section 7.3 examines the effect on sectoral energy use of including a mechanism emphasizing the existence of complementarities between technologies. In section 7.4 the effect of including a learning-by-using process for new technologies is analysed. Section 7.5 deals with the impact of different assumptions concerning the shape of the learning and diffusion curves in the model. Section 7.6 concludes.

7.2 NEMO: MODEL AND CALIBRATION

In this section we introduce NEMO by describing its main mechanisms as well as its calibration, drawing upon Koopmans et al. (1999). NEMO is a top-down energy demand model for the Dutch economy, distinguishing 20 sectors including 14 industrial sectors, services, transport, agriculture, horticulture, government and households (see Table 7.A.1 in Appendix 7.A).

Model

The model structure of NEMO is based on a nested KLEM-type production function (Berndt and Wood 1975), combining two energy inputs (electricity and fossil fuels) with capital, labour and materials, as is shown in Figure 7.1.

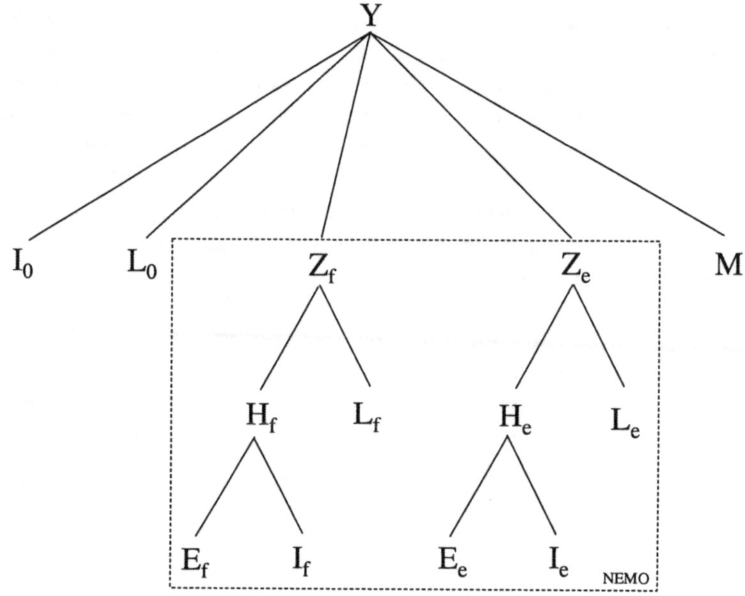

Figure 7.1. NEMO's production function structure

As can be seen from the figure, output (Y) is produced by combining capital (I_o) and labour (L_o) inputs that cannot be substituted for labour, two types of energy services (fuel[2] Z_f and electricity Z_e) and materials (M). Energy services are made up of labour that can be substituted for energy (L_f and L_e) in combination with a capital–energy bundle (H_f and H_e), the latter being a combination of capital that can be substituted for energy (I_f and I_e) and energy consumption (E_f and E_e). NEMO is used in combination with other economic models (ATHENA and STREAM) computing the volumes of production (Y) and the non-energy inputs labour (L_o), capital (I_o) and materials (M). NEMO is confined to describe only the substitution between capital, labour and energy in the production of energy services (Z_f and Z_e), as is reflected by the box in Figure 7.1.

A CES specification is used to describe substitution possibilities between capital, labour and energy. Cost minimization yields the demand equations for these production factors (see Koopmans et al. 1999). However the database used to estimate the parameters does not contain information on the levels of the capital–energy bundle (H), energy-saving capital (I) and energy use (E), but only provides information about additional capital investments necessary to save energy (see below). Therefore the factor demand equations are log (linearized) yielding constant price elasticities independent of *levels* of factor demand.[3] This implies that NEMO is a partial model, without a full representation of producer behaviour.

NEMO is a vintage model and assumes that technology is embodied in physical capital. Technological change takes the form of decreasing energy–capital ratios, that is, increasing energy efficiency. Since technological change is incorporated in new capital goods, it only becomes effective through installation of new machinery equipment. Technological change is thus directly connected with investment (and scrapping) decisions while the capital stock at any point in time consists of a range of heterogeneous vintages of different age and with different degrees of energy intensity. The model uses a so-called putty–semi–putty type of production function (Fuss 1977) to describe the substitution possibilities between the capital, labour and energy production factors in the production of energy services. This implies that production factors are substitutable both *ex ante* (at the time of investment) and *ex post* (after installation), but *ex post* substitution possibilities are more limited than *ex ante* substitution possibilities. Within the vintage structure of NEMO, *ex ante* efficiency improvements are realised through replacement investments, while *ex post* efficiency improvements can be realized in two alternative ways during the lifetime of a vintage. First, additional investments aimed at energy saving may be added to existing vintages, for example by adding end-of-pipe technologies. This is called 'retrofit investments'. Second, additional labour can be substituted for energy. This is called 'good housekeeping'. In this vintage structure, responses to a rising energy price are partly instantaneous, through *ex post* substitution (retrofit investments and good housekeeping), and partly gradual, through *ex ante* substitution (replacement investments). Hence price elasticities are larger in the long run than in the short run, and the full impact of a price change is achieved only after the capital stock has been completely renewed.

For a detailed description of the model and formal derivation of energy use we refer to Koopmans et al. (1999) and Koopmans and te Velde (2001). Here, we summarize the model by briefly discussing the main equations. Essentially, the model takes the intensity level in a base year t_0 as the numeraire and then determines the relative energy intensity level[4] of the capital stock in year t in five steps. First, the relative *ex ante* energy intensity (F^{ea}) of the vintage installed in year τ is defined as (indices for fuel and electricity are suppressed for convenience):

$$F_{\tau,t}^{ea} \equiv \frac{E_{\tau}^{ea} / Z_{\tau}}{E_{t_0} / Z_{t_0}} = e^{-\alpha(\tau - t_0)} \left(\frac{P_{E,\tau}}{P_{E,t_0}} \right)^{-\beta}, \tag{7.1}$$

where α is the vintage trend reflecting exogenous energy efficiency improvements of subsequent new replacement technologies, P_E the energy price and β the price elasticity. As in all traditional vintage models, the total amount of investment in year t is allocated to the newest vintage (see for example Salter 1960, Meijers 1994). Recall from our discussion in Chapter 3 that the very reason that old and new vintages coexist in these vintage models is that once firms have incurred the (partly) sunk investment cost, it need not be optimal to replace the existing capital goods once a new technology arrives. The same applies to NEMO's vintage structure with the implication that new technologies only gradually find their way into the economy since older vintages are only scrapped after a certain lifetime period has been exceeded (see below).

Second, the relative potential *ex post* energy intensity (F^{pr}) ('potential retrofit') in year t of the vintage installed in year τ is defined as:

$$F_{\tau,t}^{pr} \equiv \frac{E_{\tau,t}^{pr} / Z_{\tau,t}}{E_{t_0} / Z_{t_0}} = F_{\tau}^{ea} e^{-\gamma(t-\tau)} \left(\frac{P_{E,t}}{P_{E,\tau}} \right)^{-\delta}, \tag{7.2}$$

with γ the age trend reflecting energy efficiency improvements of existing vintages due to subsequent retrofit investments and δ the price elasticity. Equation (7.2) defines 'upgrading' of existing vintages, that is the extent to which the energy intensity of existing vintages can be reduced by investing in retrofit techniques. So contrary to the *ex ante* energy-intensity, which remains constant over the lifetime of a vintage, the potential *ex post* energy intensity decreases during its lifetime. This would imply that, contrary to replacement technologies, new retrofit techniques instantaneously affect the whole capital stock as soon as they arrive on the market. The observed slow diffusion of new energy-saving technologies (the energy efficiency paradox) however suggests this to be an unrealistic assumption. Therefore NEMO contains a partial adjustment process leading to a gradual diffusion of new retrofit technologies.

This is the third step, where the relative actual *ex post* energy intensity (F^{ar}) ('actual retrofit') in year t of the vintage installed in year τ is defined as:

$$F_{\tau,t}^{ar} \equiv \frac{E_{\tau,t}^{ar} / Z_{\tau,t}}{E_{t_0} / Z_{t_0}} = F_{\tau}^{ea} \qquad \text{if } t = \tau$$

$$= F_{\tau,t-1}^{ar} + \psi \left(\frac{I_t^{tot}}{K_t} \right)^{\varepsilon} \left[F_{\tau,t}^{pr} - F_{\tau,t-1}^{ar} \right] \qquad \text{if } t > \tau \text{ and } F_{\tau,t}^{pr} \leq F_{\tau,t-1}^{ar} \tag{7.3}$$

$$= F_{\tau,t-1}^{ar} \qquad \text{if otherwise}$$

where ψ and ε are adjustment parameters, I^{tot} is total investment volume and K is capital stock. So, the actual *ex post* energy intensity of a vintage installed in year τ equals the *ex ante* energy intensity at its year of installation plus the gradual adjustment to the ideal *ex post* energy intensity level during its lifetime, accounting for a retrofit energy efficiency gap. The speed of adjustment depends on the rate of replacement of the capital stock (I_t^{tot}/K_t), so that retrofit investment is influenced by the general propensity to invest.

Ex post substitution possibilities can be realised not only by *ex post* investment in retrofit technologies but also by substitution of labour for energy (good housekeeping). The fourth step is therefore to define the relative *ex post* energy intensity (F^{ep}) including good housekeeping in year t of the vintage installed in year τ, which is done as follows:

$$F_{\tau,t}^{ep} \equiv \frac{E_{\tau,t}^{ep} / Z_{\tau,t}}{E_{t_0} / Z_{t_0}} = F_{\tau,t}^{ar} \left(\frac{P_{E,t} / w_t}{P_{E,t_0} / w_{t_0}} \right)^{-\theta}, \tag{7.4}$$

with w the price of energy-saving labour and θ the price elasticity. Finally, the relative energy intensity of the total capital stock is determined by:

$$F_t = \frac{\sum_{\tau=t-a_{\max}}^{t} (1 - c_{t-\tau}) I_\tau F_{\tau,t}^{ep}}{\sum_{\tau=t-a_{\max}}^{t} (1 - c_{t-\tau}) I_\tau}, \tag{7.5}$$

where $c_{t-\tau}$ is cumulative scrapping after τ years.

Capital scrapping is assumed to be linear over a symmetric interval around the average lifetime of a vintage [a_{min}, a_{max}], according to

$$
\begin{aligned}
c_{t-\tau} &= 0 && \text{if } t-\tau < a_{\min} \\
&= \frac{(t-\tau)-a_{\min}}{a_{\max}-a_{\min}} && \text{if } a_{\min} \le t-\tau \le a_{\max} \\
&= 1 && \text{if } t-\tau > a_{\max}
\end{aligned}
\qquad (7.6)
$$

with the average lifetime calculated at the sectoral level (see below).

To illustrate the above-described mechanisms, Figure 7.2 summarizes the first four steps determining the *ex ante* and *ex post* relative energy intensity level of the capital stock for the sector paper, pulp and printing. The left part of the figure shows the relative energy intensity at time $t = 2020$ with a cross-section view of the different vintages used in 2020, with the most recent vintage at the right. It can be seen that the most recent vintage is the most energy-efficient one and that at the time of introduction its *ex post* and *ex ante* energy intensity are equal by definition. For older vintages ($t>\tau$), however, *ex post* energy intensity is lower than *ex ante* energy intensity, due to retrofit investments and good housekeeping. Moreover, the potential *ex post* energy intensity (F^{pr}) of the different vintages is slightly lower than their actual *ex post* energy intensity (F^{ar}) because of the adjustment process (equation 7.3), with the difference getting smaller over time (see further section 7.6). Finally, good housekeeping slightly lowers the actual *ex post* energy intensity level. The right part of the figure shows the development of the relative energy intensity of a vintage installed in year $\tau = 2001$ during its lifetime.

It can be seen that, by definition, the *ex ante* energy intensity level stays constant over the lifetime of a vintage: energy efficiency improvements are embodied in subsequent vintages and do not affect a vintage already installed. But the energy intensity level of an installed vintage decreases during its lifetime due to *ex post* substitution possibilities provided by retrofit investments and good housekeeping. Again, due to the adjustment process the potential *ex post* energy intensity level (F^{pr}) of the vintage is slightly lower than its actual *ex post* energy intensity level (F^{ar}) while the latter is decreased again by good housekeeping.

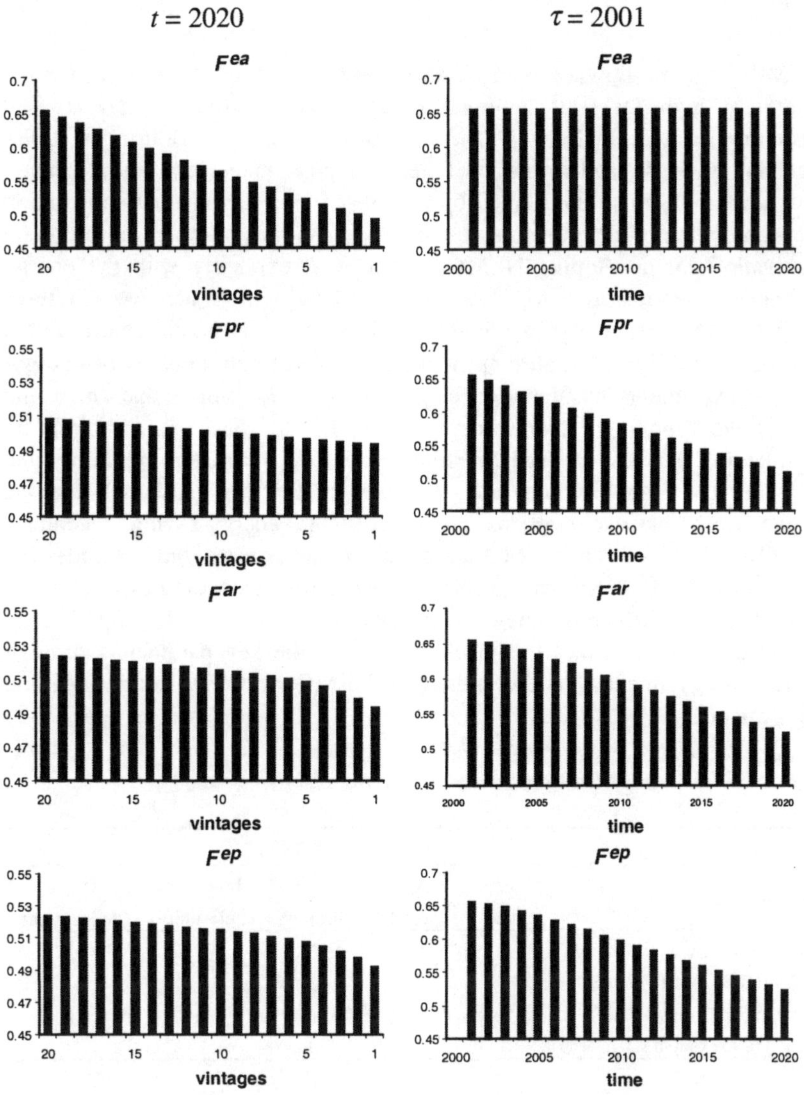

Figure 7.2 Baseline ex ante and ex post energy intensity level for the paper sector; (left) at time t= 2020 with cross-section of different vintages in use and (right) for vintage τ=2001 during its lifetime.

Calibration

NEMO is a top-down model in the sense that it is characterized by a relatively high level of aggregation and quite restrictive and stylised assumptions about technology and technological change (Böhringer 1998). Technology is defined in terms of (technological knowledge about) input–output relationships, represented by parameters and factor-input substitution elasticities in a production function structure. As noted before, an important motivation for developing NEMO was to deal explicitly with the energy efficiency paradox in a top-down model. This goal is achieved in three different ways. First, it is assumed that firms and households require a high rate of return. Based on an empirical survey about determinants of energy-saving investments by Velthuijsen (1995), NEMO assumes that firms and households apply a 15 per cent discount rate. Second, a bottom-up engineering database is used to estimate the key parameters of the model, applying the high discount rate. But even at this high discount rate the engineering database contains large profitable energy-saving potentials. Thirdly, for this reason NEMO adopts a vintage structure and includes the above described adjustment mechanism for retrofit technologies to assure a relatively slow diffusion of new technologies.

Table 7.1 summarizes the mechanisms and the key parameters through which energy intensity is determined in NEMO by trends, elasticities and adjustment.

Table 7.1 Mechanisms and determinants in NEMO

	Effect of:				
Effect on:	Vintage trend	Age trend	Price elasticity	Investment elasticity	Partial adjustment
Initial efficiency	α		β		
Theoretical ('ideal') retrofit		γ	δ		
Actual ('real') retrofit				ε	ψ
Good housekeeping			θ		

For each sector, the values of these parameters are derived from a bottom-up engineering database. This database is called ICARUS and contains information on costs and energy-saving potentials of all available energy-saving techniques in the Netherlands that were not implemented in 1990. In relying on this database, NEMO builds upon detailed information about technical, economic and environmental features of well-specified ('real') technologies. The database contains information on the annual energy-saving potential as well as investment and maintenance costs for all identified technologies. ICARUS is used to compute the profitable energy-saving potential following from the sum of the saving potentials of profitable techniques in the period 1990–2015, at constant 1990 prices. A technique is profitable if, given a discount rate of 15 per cent, the net present value of the costs is smaller than the net present value of energy savings. Using this information, the energy efficiency improvement trends α and γ can be calculated by deriving sectoral energy conservation supply curves, describing the relationship between relative energy intensity and investment costs (van Vuuren 1996). Recall that *ex post* substitution possibilities are smaller than *ex ante* substitution possibilities, implying that $\alpha > \gamma$. The price elasticities β and δ are computed by comparing the energy efficiency improvements caused by, respectively, replacement and retrofit investments at various constant real energy prices. Since it is very difficult to determine the costs of good housekeeping techniques, it is assumed that these costs are such that these good housekeeping measures were just not profitable in 1990 and hence energy efficiency improvements through good housekeeping can be achieved only after price increases. For each sector a minimum value for the good housekeeping price elasticity θ is assumed. The estimated values for the different parameters used in NEMO are given in Table 7.A.2 in Appendix 7.A. Finally, ICARUS does not contain information to estimate the adjustment parameters ψ and ε. Therefore it is assumed that $\psi = 0.9$ and $\varepsilon = 0.4$ for all sectors, such that the total adjustment parameter $\psi(I_t^{tot}/K_t)^{\varepsilon}$, implying that around 50 per cent of the profitable retrofit techniques are adopted within two years, around 75 per cent within four years, and so on. Finally, the average lifetime at the sectoral level is calculated by the weighted average of the lifetime of buildings and appliances or processes, with weights determined by energy consumption aimed at heating the building and using appliances and processes respectively. For transport and for fuel use in households however data from Statistics Netherlands (CBS) on scrapping of vehicles and houses, respectively, are used. Table 7.A.2 in Appendix 7.A provides the values of a_{min} and a_{max} for each sector. In the remainder of this chapter we present our own analyses, using NEMO.

7.3 COMPLEMENTARITIES

'Technological change is not a once-and-for-all-event but a continual process in which many different techniques coexist' (Chari and Hopenhayn 1991: 1144). Many new technologies pass through a life cycle in which they initially complement older technologies and only subsequently (and often slowly) substitute for older technologies (see for example Rosenberg 1976, 1982; Young 1993b). For this reason, in Chapter 3 it has been argued that complementarity (that is imperfect substitutability) is not so much a by-product of past investment decisions, but an essential ingredient for the process of technological change. Not only at the macro level, but also at the sector level or even at the firm level, the picture of technological change is often one of continuous investment in both old and new technologies, depending on the technology and the type of production process. Therefore, the model developed in Chapter 3 is characterized by returns to diversity of using a mix of technologies. The main reason for the existence of returns to technological diversity is that technologies often differ not only in terms of their (energy) productivity (the vertical dimension), but also with respect to other qualities (the horizontal dimension) such as, for example, input requirements, plant location, plant size, managerial and organisational skills, and so on. This 'taste for diversity' has been offered as an additional explanation for the observed slow diffusion of new technologies because it provides an incentive to invest in older vintages, even if new technologies are available that are 'better' when considered in isolation. Moreover it implies that scrapping of older vintages will be slowed down.

Although NEMO adopts a vintage structure to model slow diffusion of new technologies, it does not include this 'complementarity' explanation of the energy efficiency paradox. Recall that in NEMO, as in all traditional vintage models, the total amount of investment in year *t* is allocated to the newest vintage only (Meijers 1994). Hence the existence of complementarities between new and old technologies in NEMO is typically a by-product of past investment decisions. Nevertheless unlike many vintage models NEMO does include the possibility of investment in older vintages through what is labelled 'retrofit investment'. However these retrofit investments are represented only by *ex post* energy efficiency improvements of installed vintages (see equations 7.2 and 7.3), without any effect on the investment volume available for annual investments in best-practice replacement technologies. This is an important drawback of the model for a number of reasons. First, the assumption that upgrading of existing technologies is costless in terms of investment volume is obviously unrealistic. Second, the assumption that all investment is in best-practice technologies is found to be empirically suspect (Davies 1979, Gomulka 1976). Third, these two assumptions prevent NEMO from including an allocation mechanism by which the distribution of investment over replacement and retrofit technologies is endogenously determined. Fourth, it follows that NEMO lacks an endogenously determined scrapping date for vintages.

Ideally the allocation of investment volume to new capital goods or to the upgrading of existing ones will depend not only on the relative price and quality of the new replacement and retrofit techniques, but also on the extent to which new technologies are complementary to those already installed. In the vintage model of Chapter 3 this allocation problem is solved by deriving a demand curve for capital of specific vintages in which the relative demand for a vintage depends on its relative productivity, capital prices, energy prices and the degree of complementarity (see equation 3.12 in Chapter 3). The main reason for the omission of such a mechanism in NEMO is that information on the relative productivity (that is, energy-saving potential) and investment costs of each technology is included in the bottom-up database ICARUS, while it appears only implicitly in NEMO through the technology parameters α and γ. In doing so, NEMO assures that the technology parameters are based upon bottom-up information. But this feature comes at a certain price. The impossibility of deriving a demand curve for specific vintages implies not only the lack of an investment allocation mechanism, but also the lack of an endogenously determined scrapping date for vintages. In traditional vintage models, a downward-sloping demand curve for vintages determines the point in time at which demand for a specific vintage becomes so low that it can no longer compensate for its investment and maintenance costs (den Hartog and Tjan 1980, Malcomson 1975). In Chapter 3 we offered a somewhat different scrapping mechanism, being supply rather than demand oriented, which nevertheless builds upon the assumption that a vintage is sustained in the economy as long as demand suffices to compensate for its fixed costs.

In sum, including a mechanism in NEMO to allocate resources between replacement and retrofit technologies as well as an endogenous scrapping mechanism requires a demand curve for specific vintages, the derivation of which would imply a complete restructuring of the model, including a further integration of NEMO with the database ICARUS. This is outside the scope of this chapter because in fact this would imply a calibration of the model developed in Chapter 3. The aim of this chapter however is to include the insights obtained in Chapter 3 in an existing policy model in order to analyse its effect on estimated sectoral energy consumption and the impact of energy policies. Therefore we include a mechanism in NEMO that refines the upgrading of installed vintages in order to stress the role of complementarities between the different vintages, without a complete restructuring of the model. Additionally, we examine the potential impact of a taste for technology diversity on sectoral energy consumption by performing a sensitivity analysis with respect to scrapping ages.

Mechanism

For the purpose of refining technology diffusion in NEMO such that it captures the idea of a 'taste for diversity' of vintage, we change equation

(7.2) such that the *ex post* energy improvements of installed vintages are no longer constant during the lifetime of an existing vintage, but increase as the vintage gets older. The underlying idea is as follows. Complementarities between technologies arise from the fact that technologies differ in terms of their horizontal quality dimension (see Chapter 3). But, using a mix of vintages also requires that the differences in the vertical dimension (the energy-saving potential) should be limited. As argued in Chapter 3, modern production processes consist of many interrelated and mutually reinforcing technologies, and hence can be seen as a puzzle of a large number of interdependent and thus complementary technology pieces. The point stressed here is that the different pieces of the puzzle should fit together, although they are not the same. A higher degree of complementarity therefore requires that older vintages should be upgraded more than proportionally in order to perform well in a complex production process with many interdependencies. Hence equation (7.2) is changed into:

$$F_{\tau,t}^{pr} = F_{\tau}^{ea} e^{-\nu(t-\tau)} \left(\frac{P_{E,t}}{P_{E,\tau}} \right)^{-\delta}, \tag{7.2b}$$

with

$$\nu = \gamma(1-\alpha) \left[\frac{\varepsilon - 1}{\varepsilon} - \left[\frac{\varepsilon - 1}{\varepsilon} - 1 \right]^{1/x} (t-\tau) \right]. \tag{7.2c}$$

Equation (7.2c) yields an upward-sloping curve implying that relative energy efficiency improvements of an existing vintage increase over time. Compared to the baseline with constant *ex post* energy efficiency improvements (γ), the degree of upgrading is now initially smaller, while it becomes larger as a vintage gets older. The relationship between both curves is illustrated in Figure 7.3.A.

The point of intersection between the baseline curve and the upward sloping 'complementarity curve' ($\gamma = \nu$) is determined in equation (7.2c) by the parameter x, which we assume to be located halfway through a vintage's maximum lifetime (that is, $x = 0.5 \times a_{max}$).[5] Moreover it is shown that the stronger the complementarity effect (the lower the elasticity of substitution) the larger the difference between ν and γ. In equation (7.2c), the strength of the complementarity effect is defined by the substitution elasticity parameter ε; the smaller ε the higher the degree of complementarity and the higher the relative upgrading when a vintage gets older. As such equation (7.2c) builds implicitly upon the definition of aggregate capital as it was used in Chapter 3 (equation 7.2), with the capital stock being a composite of different vintages.[6]

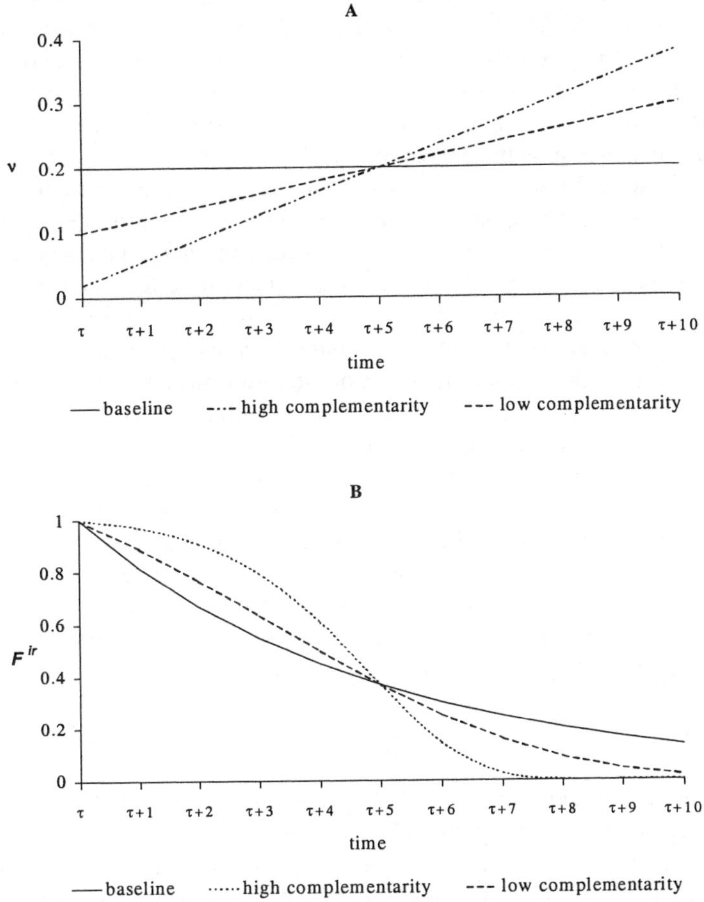

Figure 7.3 Ex post energy intensity development for vintage τ during its lifetime under different complementarity regimes; (A) shows the degree of improvement (v) and (B) shows the resulting (ideal) energy intensity level (F^{ir}) of the vintage during its lifetime.

Recall from Chapter 3 that this formulation is due to the monopolistic competition model of Dixit and Stiglitz (1977), with $1 < \varepsilon < \infty$ to assure that vintages are imperfect substitutes, implying a taste for diversity. Finally, it is assumed that upgrading of vintages during their lifetime also is a function of the technology trend parameter α: the higher α, the higher energy efficiency improvements embodied in subsequent vintages and therefore the higher the relative upgrading when a vintage gets older in order to keep up with more recent vintages. Figure 7.3b shows the resulting (potential) energy intensity level of the vintage during its lifetime. It can be seen that, as compared to the baseline, the energy intensity of a vintage decreases relatively slowly during the first half of its lifetime and relatively fast during the second half of its lifetime, depending on the degree of complementarity.

To illustrate the subsequent steps of the model in determining the energy intensity level of the capital stock, Figure 7.4 shows again for the paper, pulp and printing sector the *ex ante* and *ex post* relative energy intensity levels in the presence of the above described complementarity effect. The left part of the figure again shows the relative energy intensity at time t = 2020 with a cross-section view of the different vintages in use with the most recent vintage at the right, and the right part of the figure showing the development of the relative energy intensity of a vintage installed in year τ = 2001 during its lifetime.[7] The baseline figures in Figure 7.2 are shown in grey. Compared to the baseline, the left part of the figure shows that the *ex post* energy intensity level of recent vintages is slightly higher than the baseline, while the opposite is true for older vintages due to the fact that the degree of upgrading is relatively high when a vintage gets older. The right part of Figure 7.4 shows the implication for an individual vintage: its energy intensity level is slightly higher during the first half of its lifetime and is relatively low during the second half of its lifetime.

Results

What are the consequences of including this complementarity effect in terms of aggregate energy consumption in different sectors? Figure 7.5 shows the difference in use of electricity (Figure 7.5a) and fuel (Figure 7.5b) per sector over time as compared to the baseline in case of a standard complementarity model.[8] To analyse the *ceteris paribus* change in energy consumption, we keep energy prices and investment constant. It can be seen that the inclusion of a complementarity effect leads in general, as compared to the baseline, to a small increase in energy use in the short run, with the difference decreasing after some period of time, such that it sometimes even becomes negative in the medium run – implying a lower energy use in the medium run. This general pattern is in line with the results shown in Figures 7.3 and 7.4.

The differences between the sectors are due to a combination of differences in the estimated values for the parameters x, γ and α which, according to equation (7.2c), determine the development of *ex post* energy efficiency improvements (v). The higher the maximum lifetime a_{max}, the stronger the relative improvement of a vintage when it gets older, since we assumed $x=0.5\times a_{max}$ (see Figure 7.4).

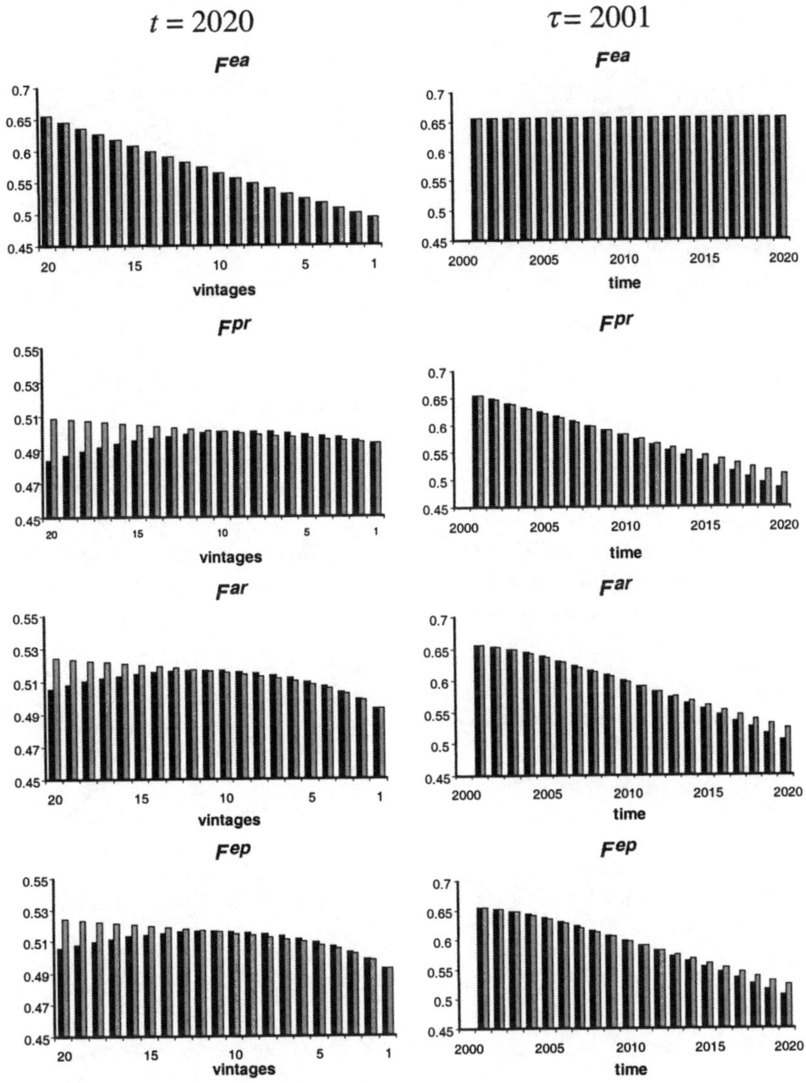

Figure 7.4 *Ex ante and ex post energy intensity levels in the Paper sector under complementarity; (left) at time t=2020 with cross-section of different vintages in use and (right) for vintage τ=2001 during its lifetime. Baseline in grey.*

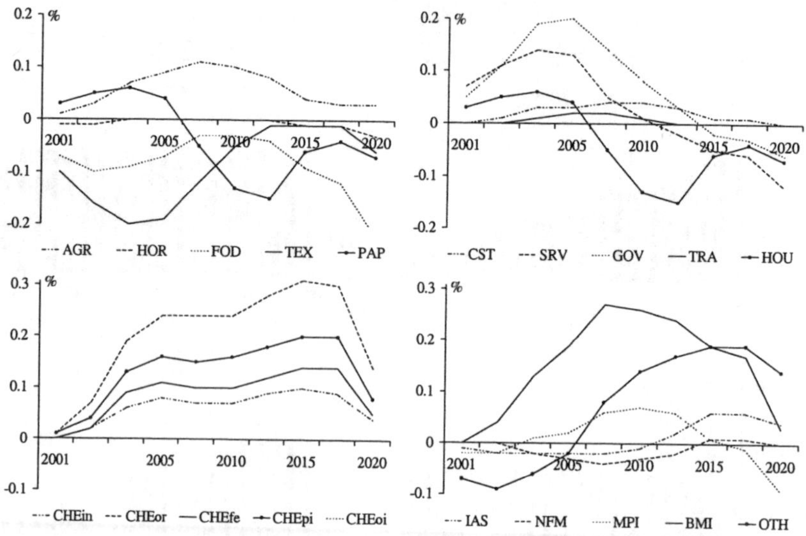

Figure 7.5a. Ceteris paribus effects of technological complementarity on electricity use (under constant energy prices and investment); % change compared to baseline

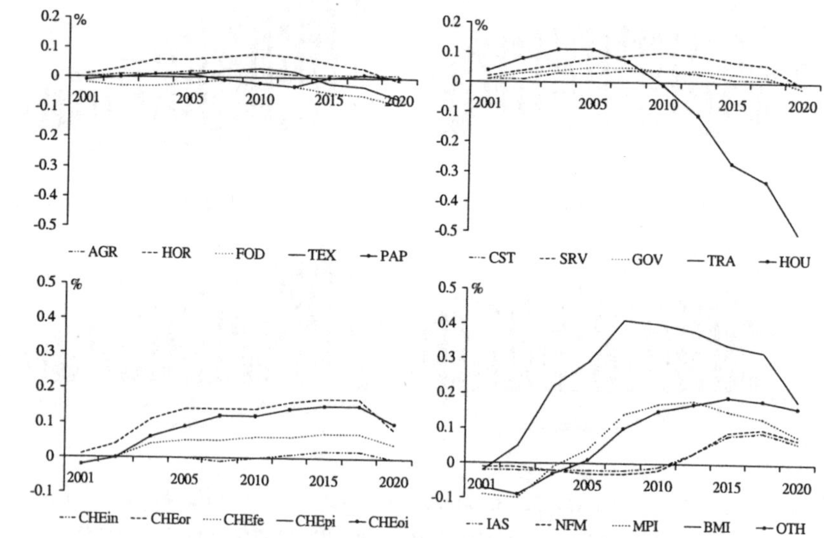

Figure 7.5b. Ceteris paribus effects of technological complementarity on fuel use (under constant energy prices and investment); % change compared to baseline

The higher the retrofit technology parameter γ, the larger the absolute difference between ν and γ, that is, the lower the initial degree of upgrading and the higher the degree of upgrading when a vintage gets older. A similar, although smaller, effect occurs in the case of a higher replacement technology parameter α. For example the relative increase in electricity consumption in the organic chemicals sector and fuel consumption in the building materials sector is due to a relatively high value of γ, which implies a relatively low initial degree of upgrading of vintages, and hence a relative increase in electricity and fuel consumption respectively. A relatively high value of γ also implies however a relatively high degree of upgrading of older vintages, inducing a relatively low energy consumption after some period of time. Which of these two effects dominates depends, as noted above, not only on γ, but also on α and even more on the maximum lifetime a_{max} determining the point of intersection between the baseline curve and the upward sloping 'complementarity curve'. For example in the paper industry the combination of a high γ with a high α for electricity use causes the latter effect to become dominant after 2006, implying lower electricity consumption as compared to the baseline. The same effect applies for example to fuel consumption by households, where the relatively low fuel consumption after 2010 is also to a considerable extent due to the high maximum lifetime of vintages (a_{max} = 50), which provides much room for improvement of older vintages before they are scrapped (see Figure 7.4).

Sectoral differences in scrapping ages are also responsible for sectoral differences in 'starting levels' of energy consumption differences in 2001, including the lower initial energy consumption in several sectors, as compared to the baseline – most notably the textile industry, food and beverage industry, the metal products sector and other industry. The long lifetime of vintages in these sectors implies that the capital stock is relatively energy-efficient due to the high degree of upgrading of older vintages. At the same time, one can see again an increase in energy consumption over time due to the introduction of new vintages that are initially upgraded at a low rate. Moreover one can see in, for example, the food industry that after some period of time the relative electricity consumption starts to decrease again, due to the high degree of upgrading as vintages become older.

Now we have analysed the *ceteris paribus* changes, we present in Table 7.2 the difference in electricity and fuel use per sector over time as compared to the baseline, for the standard complementarity case, but now using the baseline fluctuations in investment rates and energy prices (see Tables 7.A.3, 7.A.4 and 7.A.5 in Appendix 7.A). The results shown in Table 7.2 basically reproduce the patterns shown in Figure 7.5: the inclusion of a complementarity effect induces in most sectors a small increase in energy use in the short run, with the difference decreasing after some period of time, and in some sectors becoming negative – implying a lower energy use in the medium run. The change in electricity and fuel consumption ranges, respectively, from -0.08 per cent to 0.23 per cent and from –0.21 per cent to 0.13 per cent.

Table 7.2 Effects of technological complementarity on energy use

	Electricity				Fuel			
	2005	2010	2015	2020	2005	2010	2015	2020
% change as compared to baseline								
Horticulture	0.00	0.00	-0.01	-0.02	0.01	0.01	0.01	-0.01
Other Agriculture	0.04	0.04	0.02	0.02	0.00	0.02	0.03	0.01
Food & beverage industry	-0.05	-0.05	-0.12	-0.20	-0.04	0.10	0.01	-0.08
Textile industry	-0.10	-0.02	0.01	-0.01	0.00	0.05	0.08	0.09
Paper industry	0.02	-0.06	0.03	0.03	-0.03	-0.01	-0.02	0.00
Organic chemicals	0.13	0.19	0.29	0.23	0.07	0.10	0.16	0.13
Inorganic chemicals	0.04	0.06	0.10	0.08	0.02	0.02	0.02	0.03
Fertilizers	0.06	0.09	0.15	0.12	0.00	0.02	0.02	0.03
Chemical products industry	0.08	0.13	0.19	0.16	-0.01	0.06	0.04	0.05
Other chemical industry	0.08	0.13	0.19	0.16	0.00	0.06	0.05	0.05
Iron and steel	-0.01	0.00	0.04	0.03	-0.04	-0.01	0.04	0.03
Non-ferrous metals	-0.02	-0.02	0.01	0.00	-0.03	-0.02	0.04	0.03
Metal products	0.01	0.05	0.04	0.00	-0.02	-0.03	-0.01	-0.01
Building materials	0.10	0.13	0.09	-0.02	0.03	-0.05	0.04	-0.02
Other industry	-0.01	0.07	0.09	0.06	-0.03	-0.02	0.03	0.02
Construction	0.02	0.02	0.01	0.00	0.01	0.01	0.01	0.01
Services	0.07	0.00	-0.04	-0.08	0.02	-0.01	0.00	-0.02
Government	0.10	0.04	-0.01	-0.04	0.01	-0.01	0.01	-0.01
Transport	0.01	0.01	0.00	0.00	0.00	0.00	0.00	0.00
Households	0.05	0.06	0.06	0.07	0.04	-0.05	-0.14	-0.21

Next we examine the impact of energy policies in the case of complementarities by taking the model with constant energy prices and investment, and subsequently imposing a permanent increase in the energy price of 5 per cent by the year 2005, representing an energy price tax or the introduction of an emission trading scheme.[9] In NEMO, responses to a rising energy price are partly instantaneous, through *ex post* substitution possibilities (retrofit investments and good housekeeping), and partly gradual, through *ex ante* substitution possibilities (replacement investments). In Table 7.3 we compare the impact of this 'policy' in the baseline model with the impact in the complementarity model, again in terms of changes in electricity and fuel use per sector, for the standard complementarity case.

Table 7.3 Effects of a 5% increase in energy price in 2005 on energy use, under technological complementarity

	Electricity				Fuel			
	2005	2010	2015	2020	2005	2010	2015	2020
% change as compared to baseline								
Horticulture	0.00	0.00	0.00	0.00	0.00	0.00	0.00	0.00
Other Agriculture	0.00	0.00	0.00	0.00	0.00	0.00	0.00	0.00
Food & beverage industry	0.00	-1.72	0.00	0.00	0.00	0.00	0.94	0.00
Textile industry	0.00	0.00	0.00	0.00	4.35	7.89	2.94	0.86
Paper industry	0.00	0.00	0.00	0.00	0.00	1.25	0.81	0.62
Organic chemicals	0.00	0.00	0.00	0.00	0.00	0.00	0.00	0.00
Inorganic chemicals	0.00	0.00	0.00	0.00	0.00	0.00	0.00	0.00
Fertilizers	0.00	0.00	0.00	0.00	0.00	0.00	0.00	0.00
Chemical products industry	0.00	0.00	0.00	1.37	0.00	0.00	0.00	0.00
Other chemical industry	0.00	0.00	0.00	1.37	0.00	0.00	0.00	0.00
Iron and steel	0.00	0.00	0.00	0.00	0.00	0.00	0.00	0.00
Non-ferrous metals	0.00	0.00	0.00	0.00	0.00	0.00	0.00	0.00
Metal products	0.00	0.00	0.00	2.17	-3.23	0.00	0.00	0.00
Building materials	0.00	0.00	0.00	0.00	0.00	0.00	0.81	0.00
Other industry	0.00	0.00	0.00	0.00	0.00	0.00	0.00	0.00
Construction	0.00	0.00	0.00	0.00	0.00	0.00	0.00	0.00
Services	0.00	0.00	1.09	0.00	0.00	0.00	0.00	0.00
Government	0.00	0.00	0.00	0.00	0.00	0.00	0.00	0.00
Transport	0.00	0.00	0.00	0.00	0.00	0.00	0.00	0.00
Households	0.00	0.00	0.87	0.00	0.00	0.00	0.00	0.00

It can be seen that in most sectors the inclusion of a complementarity effect does not alter the impact of an energy tax. In some sectors however there is a small positive change in impact of an energy price increase, which is mainly attributed to relatively high sectoral price elasticities (for example in the textiles and paper sectors). The negative change in impact of a higher energy price on, respectively, electricity and fuel consumption in the food and metal products industry is due to a combination of a relatively strong complementarity effect (because of a relatively high value of γ) and a relatively long lifetime of vintages. Recall that in NEMO *ex post* substitution possibilities are more limited than *ex ante* substitution possibilities, due to its putty–semi–putty type of production function. Hence, the full impact of a price change is achieved only after the capital stock has been completely renewed, which takes a long time in sectors which exhibit a relatively long lifetime of vintages. Moreover the long lifetime of vintages leaves much room for *ex post* efficiency improvements, in particular if the complementarity effect is strong. Altogether, this will reduce the impact of an increase in energy price on energy consumption.

In Chapter 3, the imposition of an energy tax led to a reduction in the amount of energy used and a slowdown of the diffusion of new vintages. The first effect was due to the assumption of no substitution possibilities between energy and capital. Under this assumption, an increasing energy price increases the costs of the energy–capital composite, which induces a lower demand for vintage capital. The latter effect was caused by the set-up of the model in which the scrapping condition was determined at the supply side: a higher energy price enables the producer of a vintage to set a higher mark-up because of the lower weight of capital costs in determining a firm's investment decision. Hence the producer needs less demand to sustain a vintage, more vintages are sustained and the adoption of new vintages slows down. NEMO neither assumes a constant energy–capital ratio nor does it include an endogenous scrapping mechanism. In Chapter 3 it was noted that it is more realistic to allow for (exogenous) improvement of energy efficiency over time, and in that case an energy tax would lead to decreasing energy consumption. That is what we see here. Moreover, it was noted in Chapter 3 that the total effect of an energy tax was ambiguous, depending on a rebound effect: energy consumption might increase as increasing energy efficiency makes the energy–capital composite cheaper.

Above, it has been argued that a taste for technological diversity has two main implications. First, it provides an incentive to invest in older vintages, even if new technologies are available that are 'better' when considered in isolation. Second, it implies that scrapping of older vintages will slow down. The first effect is captured above by assuming that this implies an increasing degree of upgrading of existing technologies as their age increases. Because of the impossibility of deriving an endogenous demand curve for vintage capital, we are not able to include an endogenous scrapping mechanism in NEMO. However in order to examine the potential impact of a delay in scrapping on sectoral energy consumption, we calculated the difference in electricity and fuel use, as compared to the baseline, under an increase of the average lifetime of five years. The results are presented in Table 7.4. From the table it can be concluded that the effect differs substantially among sectors, with the change in electricity and fuel consumption varying, respectively, between –0.45 per cent and 2.53 per cent and between –0.55 per cent and 0.12 per cent.

Table 7.4 *Effects of a 5-year increase of scrapping date on energy use, under a complementarity effect*

	Electricity				Fuel			
	2005	2010	2015	2020	2005	2010	2015	2020
% change as compared to baseline								
Horticulture	0.42	1.19	1.73	2.53	0.35	0.56	0.43	0.10
Other Agriculture	0.74	1.20	1.29	0.88	0.47	0.41	0.03	-0.05
Food & beverage industry	0.07	0.07	0.11	0.14	-0.05	-0.30	-0.11	0.05
Textile industry	0.16	0.04	-0.51	-0.65	0.02	-0.14	-0.26	-0.26
Paper industry	0.22	0.35	0.17	0.29	0.06	0.82	0.93	-0.08
Organic chemicals	0.00	0.12	0.02	0.17	0.17	0.35	0.28	0.39
Inorganic chemicals	-0.01	0.07	0.04	0.07	-0.05	-0.07	-0.08	-0.09
Fertilizers	-0.37	-0.37	-0.39	-0.41	0.00	0.04	0.00	-0.02
Chemical products industry	0.02	0.18	0.13	0.20	-0.03	-0.01	-0.09	-0.06
Other chemical industry	0.02	0.18	0.13	0.20	-0.02	-0.01	-0.09	-0.07
Iron and steel	0.01	0.02	-0.03	0.09	-0.39	-0.33	-0.44	-0.55
Non-ferrous metals	-0.55	-0.30	-0.25	-0.45	0.05	0.01	-0.07	0.02
Metal products	0.02	0.01	0.06	0.10	-0.01	-0.07	-0.06	-0.01
Building materials	-0.01	0.12	0.19	0.27	-0.10	0.03	0.09	-0.07
Other industry	-0.04	-0.05	-0.02	0.06	-0.01	-0.07	-0.09	-0.02
Construction	-0.22	-0.40	-0.34	-0.10	0.17	0.29	0.28	0.10
Services	0.35	0.61	0.69	0.67	0.05	0.10	0.12	0.12
Government	0.40	0.83	1.07	0.97	0.02	-0.01	-0.05	-0.07
Transport	0.21	0.07	1.03	1.22	0.64	1.50	-0.45	-0.15
Households	1.01	1.85	1.82	1.29	-0.08	-0.11	-0.11	-0.16

This result underlines the difference in the way in which complementarity has been modelled in Chapter 3 – where an increasing number of vintages implied a slowdown in adoption of new more efficient vintages and hence an increase in energy consumption. Here an increasing number of vintages implies an increasing upgrading of older vintages and not so much a slowdown of new technologies. Hence in some sectors the overall effect on energy consumption is negative, that is, a relative low energy consumption under complementarity as compared to the baseline.

Sensitivity Analysis

So far we have assumed the substitution parameter ε to equal 3, determining the degree of complementarity, which we labelled the standard complementarity case. We subsequently examined the sensitivity of the obtained results for the degree of complementarity. For this purpose we defined a high-complementarity version with $\varepsilon = 1.5$ and a low complementarity version with $\varepsilon = 15$.[10] The sensitivity analysis revealed that the impact of a complementarity effect on electricity consumption ranges from –0.36 per cent to 0.47 per cent under high-complementarity and from –0.05 per cent to 0.05 per cent under low complementarity. The impact on fuel consumption ranges from –0.4 per cent to 0.25 per cent under high-complementarity and from –0.05 per cent to 0.03 per cent under low-complementarity. For sectoral details we refer to Table 7.B.1 in Appendix 7.B, which presents the difference in electricity and fuel consumption per sector, as compared to the baseline, for both the high and low complementarity case.

7.4 LEARNING-BY-USING FOR REPLACEMENT TECHNOLOGIES

Throughout this volume we have put forward the interaction between technology diffusion and knowledge accumulation as an important explanation to the question of why diffusion of energy-saving technologies is often slow. In Chapters 3 and 4 we have highlighted various types of learning processes and we have shown that these various forms of knowledge accumulation bear important implications for diffusion processes. The basic idea is that new technologies improve due to learning processes, as a result of which firms experience productivity gains, and hence decreasing costs of production. This process has been documented extensively in the empirical literature on learning or experience curves, describing a negative relationship between the costs of new technologies or products on the one hand and their cumulative output or installed capacity on the other hand (see for example Hirsch 1952, Grübler et al. 1999, Argote and Epple 1990, OECD/IEA 2000).

Much can be said about the behavioural processes that give rise to learning processes, and hence drive the learning curves (see for example Adler and Clark 1991, Usher 1954, Slaughter 1993). However essentially two main sources for learning can be distinguished: those which are internal and those which are external to a firm. External sources for learning refer for example to processes of information gathering or the (exogenous) arrival of improved versions of a technology (see Chapter 4). An important internal source of learning is to build up expertise by using technologies (see Chapter 3), a process we refer to as learning-by-using (following Rosenberg 1982). Concerning the relationship between the different sources for learning, 'complementarity between sources internal and external to the firm is the general property' (Dosi 1997: 1532) – a point of view which has been subscribed to by many authors (see for example Cohen and Levinthal 1989, Freeman 1994, Malerba 1992, Slaughter 1993).

Despite the important implications of learning processes for adoption and diffusion processes, NEMO does not explicitly include a learning process. Recall that there is a continuous arrival of new replacement and retrofit technologies, with an inherent improvement of the energy-saving potential that is external to the adopters. Whereas new retrofit technologies diffuse slowly due to the adjustment process described by equation (7.3), new replacement technologies are adopted instantaneously since they are embodied in the newest vintage and the total amount of investment in year t is allocated to the newest vintage. Given the absence of a learning effect, the latter implies that the *ex ante* energy-saving potential of a new vintage is realized at the moment of installation. However technologies always embody a certain degree of tacit knowledge that is firm-specific, local and cumulative ('expertise'). Hence the full potential of new technologies is often only realized after some period of time with users of a new technology learning over time to increase their productivity as they gain expertise in how better to use the technology. Moreover not only external but also internal learning processes affect diffusion processes, since 'imperfect understanding and imperfect, path-dependent, learning entails persistent heterogeneity among agents, even when facing identical information and identical notional opportunities' (Dosi 1997: 1531). In principle the speed of diffusion depends negatively on learning-by-using. Since users of a technology explore the productivity potential of a technology by using it, they build up expertise which will be (temporarily) lost if they switch to a new technology, thereby creating a barrier for adoption of new technologies. This effect has been included in the vintage model developed in Chapter 3. In this section, we build upon this idea and drop the unrealistic assumption in NEMO that the *ex ante* energy-saving potential of a new vintage is realized at the moment of installation, and replace it by a learning-by-using mechanism through which the energy-saving potential of a vintage is explored during its lifetime.

Mechanism

Similar to Chapter 3, we include in NEMO a learning-by-using mechanism by allowing for new technologies to be initially inferior to more mature technologies (along the vertical dimension). As a result the maximum vintage-specific energy-saving potential is realized only after some period of time. As in Chapter 3, this implies that a switch of technologies temporarily reduces expertise and thereby the energy efficiency level of the capital stock. This effect is not only in accordance with broad historical evidence for many technologies (see for example Mokyr 1990, Rosenberg 1982, Young 1993a) but can also be observed – as for example in the chemical industry where the installation of new energy-saving process technologies interrupts the production process, thereby temporarily reducing the overall energy efficiency performance due to the fact that it takes some time to tune and optimize the production process again. Therefore this mechanism introduces into NEMO an additional explanation for the energy efficiency paradox, since the loss of expertise and the resulting temporary reduction in the productivity of the capital stock implies that the actual energy intensity level of the capital stock is lower than one would expect on the basis of a laboratory type of assessment of the energy-saving potential of new technologies, as represented in the engineering database ICARUS.

We have chosen to include a learning-by-using mechanism analogous to the adjustment process for retrofit technologies (equation 7.3). We change equation (7.1) such that

$$
\begin{aligned}
F_{\tau,t}^{ea\_l} &= P_\tau \cdot F_{\tau,t}^{ea} && \text{if } t=\tau \qquad (7.1b)\\
&= F_{\tau,t-1}^{ea\_l} + \psi\left(\frac{I_\tau}{K_t}\right)^{\varepsilon}\left[F_{\tau,t}^{ea} - F_{\tau,t-1}^{ea\_l}\right] && \text{if } t>\tau \text{ and } F_{\tau,t}^{ea} \le F_{\tau,t-1}^{ea\_l}\\
&= F_{\tau,t-1}^{ea} && \text{if otherwise}
\end{aligned}
$$

where $F_\tau^{ea\_l}$ is the relative *ex ante* energy of a vintage installed in year τ subject to learning-by-using and P_τ the vintage-specific learning potential to be explored through this learning-by-using. In other words, at its year of installation the *ex ante* energy intensity of a vintage equals a potential level of energy intensity which is realised over time such that it converges gradually to the baseline *ex ante* energy intensity level as defined in equation (7.1). The learning rate is assumed to depend on the relative amount of vintage specific investment (I_τ/K_t). As in the baseline model we assume that $\psi = 0.9$ and $\varepsilon = 0.4$ for all sectors, such that the learning parameter $\psi(I_\tau/K)^\varepsilon \approx 0.3$, implying that around 50 per cent of the learning potential is realized within two years, around 75 per cent within four years, and so on. Below we present a sensitivity analysis for this learning parameter, distinguishing between 'fast learning' and 'slow learning'. An additional assumption is to be made concerning the learning potential P_τ. We assume $P_\tau = 1.10$ for all sectors, implying a learning potential of 10 per cent. Below we present a sensitivity analysis for the learning potential, distinguishing between a learning potential of, respectively, 5 per cent and 10 per cent. Finally, model consistency requires that the expression $F_\tau^{ea\_l}$ is substituted for F_τ^{ea} in equations (7.2) and (7.3), to assure that *ex post* energy efficiency improvements apply to the actual instead of the potential *ex ante* energy intensity level of a vintage.

Again, we illustrate the mechanism by showing the sequential steps of the model in determining the *ex ante* and *ex post* energy intensity level of the capital stock in the paper, pulp and printing sector; see Figure 7.6. The left part of the figure again shows the relative energy intensity at time $t = 2020$ with a cross-section view of the different vintages in use with the most recent vintage at the right, and the right part of the figure showing the development of the relative energy intensity of a vintage installed in year $\tau = 2001$ during its lifetime. The baseline figures as shown in Figure 7.2 are again included in grey.

Compared to the baseline, the left part of the figure shows that in case of learning-by-using the most recent vintage is no longer the most efficient one. The most energy-efficient vintage is of intermediate age, due to the fact that the energy-saving potential inherent to the newest vintage is realized only after some period of time because of a learning-by-using effect. The right part of the figure shows again the implication for an individual vintage: initially the *ex ante* energy intensity level is higher than in the baseline model and it is no longer constant during its lifetime, but converging to its (baseline) *ex ante* energy-saving potential. Of course, this implies that the *ex post* energy intensity level is also initially higher than the baseline while the difference declines over time. As stated before, the fact that the newest vintage is not the most productive one, due to a learning-by-using effect, has already been offered in Chapter 3 as an explanation for a slow diffusion of technologies.

Results

What are the consequences of including this learning effect for the estimated aggregate energy consumption in different sectors? Figure 7.7 shows the difference in use of electricity (Figure 7.7a) and fuel (Figure 7.7.b) per sector over time as compared to the baseline, under constant energy prices and investment, and assuming a learning potential of 10 per cent. It can be seen that including a learning-by-using effect leads to an increase in energy use of about 2 per cent over a 20-year period. The differences between sectors are mainly to be explained from differences in average lifetime of vintages. The longer the lifetime of a vintage, the larger the relative importance of *ex post* energy efficiency improvements through retrofit investments and good housekeeping and the smaller the impact of a delay in *ex ante* energy efficiency improvements due to learning-by-using.

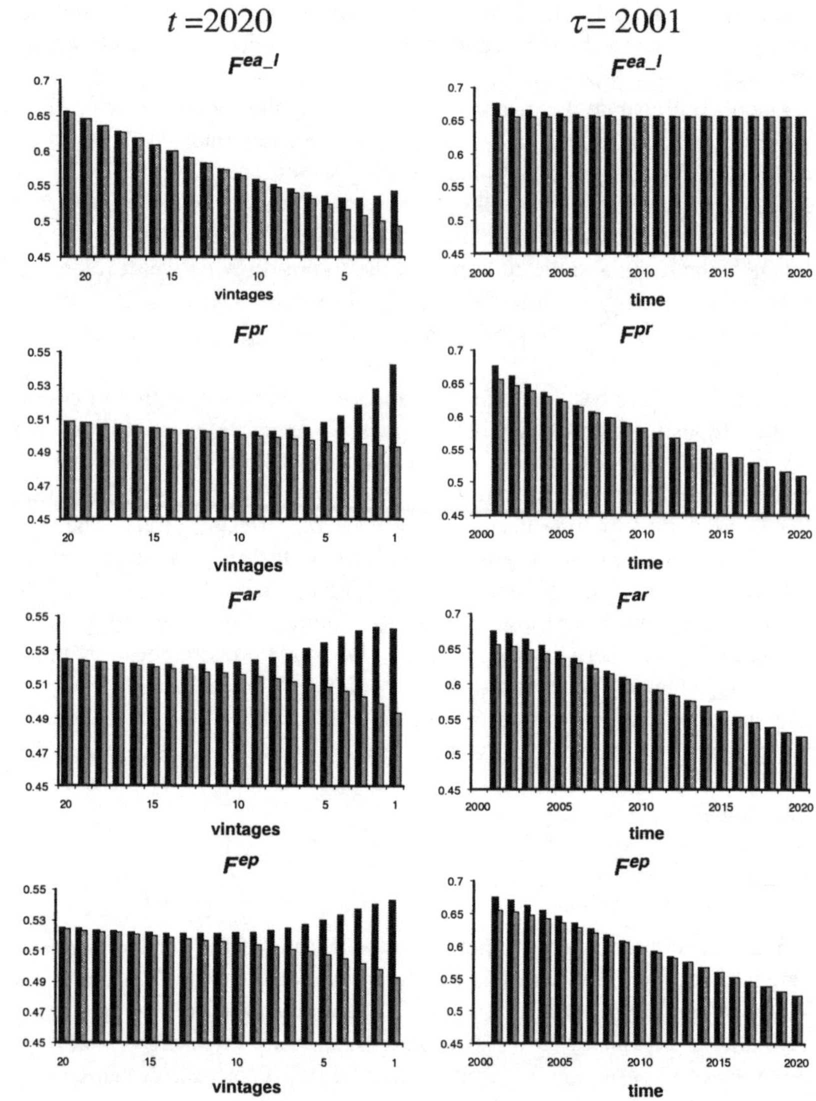

Figure 7.6 *Ex ante and ex post energy intensity in the paper sector under replacement learning; (left) at time t=2020 with cross-section of different vintages in use and (right) for vintage τ=2001 during its lifetime. Baseline in grey.*

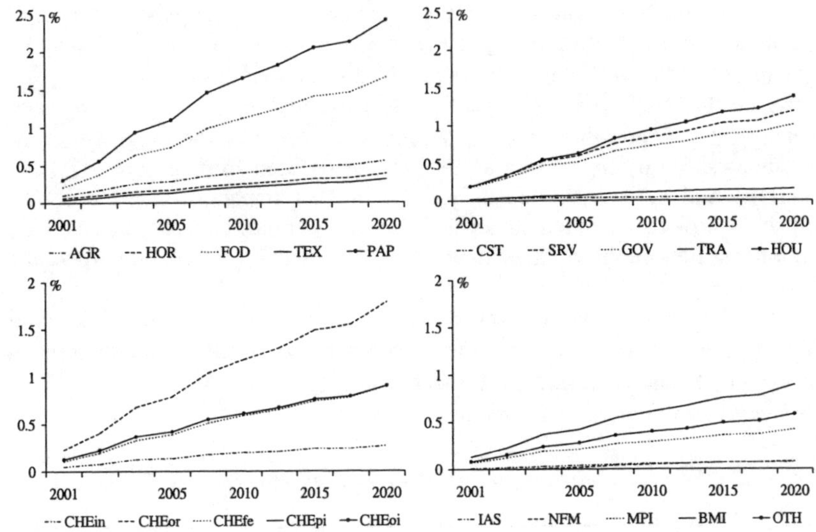

Figure 7.7a Ceteris paribus effects of a learning-by-using effect on electricity use (under constant energy prices and investment); % change compared to baseline

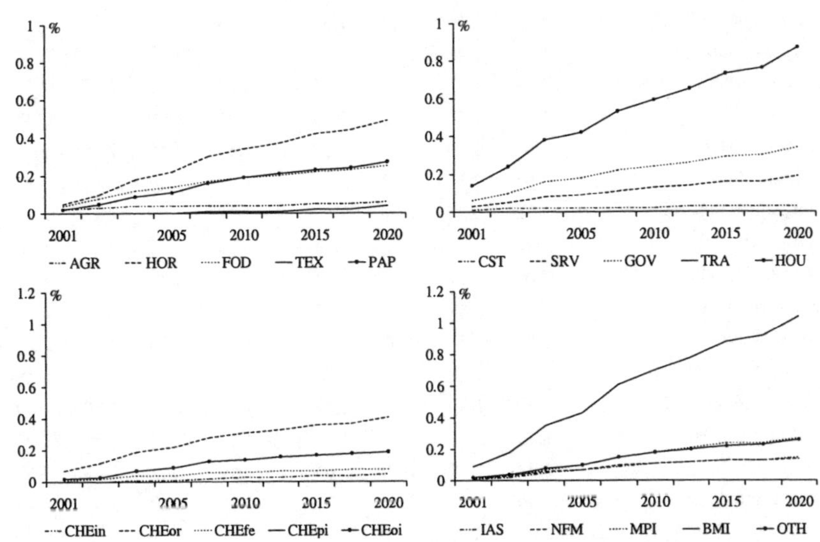

Figure 7.7b Ceteris paribus effects of a learning-by-using effect on fuel use (under constant energy prices and investment); % change compared to baseline

Figure 7.7a shows that for electricity consumption the highest impact of the inclusion of a learning-by-using effort is to be found in transport, agriculture and households – sectors with the shortest lifetime of vintages – while the lowest impact is to be found in the textile, food and beverages, metal products and other industry sectors – sectors with the longest lifetime of vintages. With respect to fuel use, it can be seen from Figure 7.7b that transport and agriculture again face the highest impact, while the lowest impact is now to be found in services and household, because they contain the longest lifetime of vintages in terms of fuel (see Table 7.A.2 in Appendix 7.A).

Table 7.5 presents the difference in electricity and fuel use per sector over time as compared to the baseline, assuming the baseline fluctuations in energy prices and investment rates (see Table 7.A.3–7.A.5 in Appendix 7.A) for a learning potential of 10 per cent.

Table 7.5 Effects of learning-by-using on energy use

	Electricity				Fuel			
	2005	2010	2015	2020	2005	2010	2015	2020
% change as compared to baseline								
Horticulture	1.36	1.76	1.84	1.86	1.71	2.22	2.31	2.31
Other Agriculture	1.97	2.46	2.57	2.62	2.16	2.98	2.83	2.63
Food & beverage industry	1.22	1.75	1.96	2.05	1.31	2.37	2.47	2.33
Textile industry	1.13	1.71	1.82	1.74	1.45	2.92	3.25	3.13
Paper industry	1.67	2.19	2.28	2.17	1.60	2.45	2.48	2.29
Organic chemicals	2.03	2.38	2.35	2.26	2.09	2.44	2.41	2.31
Inorganic chemicals	2.03	2.37	2.34	2.25	2.19	2.58	2.45	2.29
Fertilizers	1.96	2.30	2.28	2.18	2.09	2.61	2.43	2.28
Chemical products industry	2.02	2.37	2.34	2.25	2.18	2.93	2.56	2.31
Other chemical industry	2.02	2.37	2.34	2.25	2.19	2.96	2.58	2.32
Iron and steel	1.79	2.28	2.27	2.07	1.81	2.30	2.27	2.05
Non-ferrous metals	1.73	2.22	2.22	2.02	1.80	2.31	2.30	2.10
Metal products	1.58	1.89	1.80	1.64	1.68	2.12	1.89	1.69
Building materials	1.75	2.06	2.05	1.97	1.84	2.37	2.21	2.08
Other industry	1.44	1.79	1.79	1.65	1.50	1.85	1.82	1.67
Construction	1.97	2.28	2.30	2.32	2.09	2.39	2.41	2.43
Services	1.72	2.17	2.30	2.32	1.37	1.71	1.64	1.50
Government	1.75	2.21	2.35	2.39	1.36	1.75	1.79	1.73
Transport	2.22	2.50	2.61	2.59	2.34	2.76	2.86	2.84
Households	1.72	2.33	2.51	2.56	1.21	1.68	1.81	1.85

These results resemble the patterns shown in Figure 7.7, with an increase in electricity consumption ranging from 1.64 per cent to 2.62 per cent while the impact on fuel consumption ranges from 1.5 per cent to 3.13 per cent.

Next we examine again the impact of energy policies by comparing their impact in the baseline model with their impact in the model including learning-by-using for replacement technologies. Again, we simulate energy policies by imposing a permanent increase in the energy price of 5 percent by the year 2005. Table 7.6 presents the difference of policy impact between the two model versions, in terms of changes in electricity and fuel use per sector.

Table 7.6 *Effects of a 5% increase in energy price in 2005 on energy use, under learning-by-using*

	Electricity				Fuel			
	2005	2010	2015	2020	2005	2010	2015	2020
% change as compared to baseline								
Horticulture	0.00	0.00	0.00	0.00	0.00	2.13	1.56	0.00
Other Agriculture	4.00	1.52	1.04	0.00	0.00	1.35	0.00	0.00
Food & beverage industry	0.00	0.00	1.61	0.00	3.45	1.19	1.89	0.81
Textile industry	0.00	0.00	0.00	2.56	0.00	1.32	0.98	0.86
Paper industry	0.00	0.00	0.00	0.00	4.17	3.75	2.44	1.23
Organic chemicals	0.00	0.00	0.00	0.00	0.00	0.00	0.00	0.00
Inorganic chemicals	0.00	0.00	0.00	0.00	0.00	0.00	0.00	0.00
Fertilizers	0.00	0.00	0.00	0.00	0.00	0.00	0.00	0.00
Chemical products industry	0.00	0.00	0.00	1.37	0.00	0.00	0.00	0.00
Other chemical industry	0.00	0.00	0.00	1.37	0.00	0.00	0.00	0.00
Iron and steel	0.00	0.00	0.00	0.00	0.00	0.00	0.00	0.00
Non-ferrous metals	0.00	0.00	2.50	2.27	0.00	3.33	0.00	0.00
Metal products	0.00	0.00	0.00	2.17	0.00	0.00	0.00	1.33
Building materials	0.00	0.00	0.00	0.00	0.00	0.99	0.81	0.71
Other industry	0.00	0.00	0.00	0.00	0.00	0.00	0.00	0.00
Construction	0.00	0.00	0.00	0.00	5.88	2.50	0.00	0.00
Services	0.00	0.00	1.09	0.00	0.00	1.79	0.00	1.39
Government	0.00	1.09	0.00	0.00	0.00	0.00	1.43	0.00
Transport	0.00	0.00	0.00	0.00	2.50	2.70	0.54	0.00
Households	0.00	1.25	1.74	0.69	0.00	0.70	0.63	0.00

The table reveals that some sectors exhibit a small positive change in the impact of an increase in energy price on energy consumption, while there is no change in policy impact in most sectors. The increased impact of an energy tax policy in some sectors in the learning-by-using case is again largely explained by a relatively high price elasticity, for example for electricity consumption in agriculture, government and households and for fuel consumption in food, paper, iron and steel, building materials and transport. Moreover the relatively short lifetime of vintages in construction and transport contributes to a relatively large increase in impact of an energy tax, since it takes a relatively short period of time in these sectors before the capital stock has been completely renewed, and hence the full impact of a price change is achieved.

Sensitivity Analysis

As noted above, we have made two additional assumptions with the introduction of a learning-by-using effect. The first concerned the speed of learning and the second concerned the size of the learning potential. Therefore we performed a sensitivity analysis for both assumptions. With respect to the speed of learning, we assumed so far $\psi = 0.9$ and $\varepsilon = 0.4$ in equation (7.1b) for all sectors, such that the parameter defining the speed of learning $\psi(I_\tau/K)^\varepsilon \approx 0.3$, implying that around 50 per cent of the learning potential is realized within two years, around 75 per cent within four years, and so on. By means of a sensitivity analysis we examine the effect of this speed of learning on energy use and the impact of energy policies. We did so by distinguishing between a 'fast' and a 'slow' speed of learning. We defined fast learning by assuming that $\psi = 1.8$ and $\varepsilon = 0.4$ for all sectors, such that the speed of learning $\psi(I_\tau/K)^\varepsilon \approx 0.6$, implying that around 75 per cent of the learning potential is realised within two years, around 90 per cent within four years, and so on. We defined slow learning by assuming that $\psi = 0.6$ and $\varepsilon = 0.4$ for all sectors, such that the speed of learning $\psi(I_\tau/K)^\varepsilon \approx 0.2$, implying that almost 50 per cent of the learning potential is realised within two years, but only 65 per cent within four years, and so on. The sensitivity analyses revealed that NEMO is rather sensitive to the assumption concerning the speed of learning. Whereas the introduction of learning-by-using under a baseline speed of learning led to an increase in energy use of about 2 per cent over a twenty-year period, assuming a fast learning speed reduces the difference to an increase of only 0.5 per cent on average, while assuming a slow learning speed increases the difference to 3.5 per cent on average. We refer to Table 7.B.2 in Appendix 7.B for the calculated differences in electricity and fuel use per sector over time as compared to the baseline for both fast and slow learning, assuming baseline fluctuations in energy prices and investment rates.

Subsequently we performed a sensitivity analysis with respect to the learning potential. So far we have assumed the learning potential to be 10 per cent. For the sensitivity analysis we assumed a learning potential of 5 per cent and 20 per cent. Whereas a learning potential of 10 per cent led to an average increase in energy consumption of about 2 per cent (see Table 7.7), this increase is on average 1 per cent if the learning potential is assumed to be 5 per cent and 4 per cent if the learning potential is assumed to be 20 per cent. In other words, every increase (decrease) in the learning potential by 5 per cent leads to an increase (decrease) of the energy consumption by on average 1 per cent. We refer to Table 7.B.3 in Appendix 7.B for the calculated differences in electricity and fuel use per sector over time as compared to the baseline, under the baseline fluctuations in energy prices and investment rates and for a learning potential of 5 per cent and 20 per cent.

7.5 THE SHAPE OF LEARNING AND DIFFUSION CURVES

An appropriate specification of adoption and diffusion processes in economic models asks for appropriate functional forms to describe these processes. In the literature a considerable number of alternative functional forms have been developed to describe learning and diffusion processes, depending on the factors that are assumed to drive these processes (see for example Lekvall and Wahlbin 1973, Zettelmeyer and Stoneman 1993, Stoneman 2001). In NEMO, the adjustment process for retrofit technologies (equation 7.3) as well as the above introduced learning-by-using effect for replacement technologies (equation 7.1b) are assumed to follow a modified exponential curve. Learning and diffusion processes are however at least as widely analysed by using an S-shaped curve. As a matter of fact, both functional forms have received substantial empirical support (Lekvall and Wahlbin 1973). Therefore in this section we will examine the effect of the shape of the learning and diffusion curves on sectoral energy use and the impact of policies, distinguishing between an S-shaped and a modified exponential form. We start with a brief description of these two basic functional forms.

In terms of a learning process, a modified exponential curve suggests that efficiency improvements are initially relatively large, but slow down over time, while learning is ultimately bounded; that is, there is a vintage-specific maximum efficiency level. This formulation captures the available empirical evidence that productivity increases at a relatively fast rate just after a technology is introduced, slows down at later stages and levels off when the technology matures (Grübler et al., 1999). In terms of a diffusion process, a modified exponential curve implies that the speed of diffusion is initially fast while levelling off as the saturation level is approached. This type of curvature has been used in Chapter 3 to describe a learning-by-using process and is used in NEMO to describe the adjustment process for retrofit technologies. The S-shaped or sigmoid curve is a well-known result of the early sociological and economic literature on technology diffusion and has been widely used ever since, in particular also in marketing studies. The classic interpretation stems from the epidemic diffusion models in which the likelihood of adoption ('infection') increases as more firms have adopted the technology ('being infected'), with the speed of adoption ('infectiousness') depending on financial characteristics (profitability and costs) and the saturation of the market ('share of the population being infected'). Although widely used, the S-shaped curve has also been widely criticized for its lack of behavioral explanation. Nevertheless 'it is also significant that in spite of the criticisms that use of the S-shaped logistic curve lacks adequate grounding in either microeconomic or communication theory, efforts to dislodge it have been unsuccessful. It has remained remarkably robust as a description of both the diffusion and the substitution processes' (Ruttan 2001: 161). Applying the S-curve to learning processes implies that the efficiency improvements depend on the cumulative experience with a particular technology, such that efficiency improvements are initially low due to the fact that only a small stock of knowledge is available, while improvements are increasing as experience cumulates in order to level off when a technology reaches its maximum efficiency level. In the model developed in Chapter 4, a microeconomic motivation for an S-shaped curve has been offered in terms of an external learning process. Assuming heterogeneous firms, we assumed potential adopters to benefit from the experience of early adopters who provide feedback to the producer, resulting in improved versions of the technology. Due to the increased probability of duplication, learning occurred at a decreasing rate, yielding an S-shaped diffusion pattern (see Figure 4.2 in Chapter 4).

As noted above, the choice of the functional form used to describe the learning and diffusion process should depend on the factors that are assumed to drive these processes. In general, lack of empirical information makes it difficult to decide which curve is most appropriate (Stoneman 1983: 76). Moreover, as already stated in the previous section, the interpretation of the adjustment process for retrofit technologies is rather unclear in itself. Most likely the makers of NEMO had a traditional 'information gathering' interpretation in mind when formulating the adjustment process for retrofit technologies, assuming that each sector consists of a number of heterogeneous firms that differ among each other in terms of (access to) information about the new technology. At least, they argue (in a footnote) with respect to this adjustment mechanism for retrofit technologies, that 'the underlying assumption is that firms pay more attention to new vintages than to existing vintages. As a consequence, firms have imperfect information with respect to investments in existing vintages' (Koopmans et al. 1999: 23). Moreover, in equation (7.3) it is assumed that the speed of adjustment depends on the rate of replacement of the capital stock, so that retrofit investment is influenced by the general propensity to invest.

As such, the adjustment process can at best be characterized as a diffusion process, very similar to the classical technology diffusion studies (Davies 1979, Griliches 1957, Mansfield 1968, Stoneman 1983). In a similar vein, one can also interpret the adjustment process as a diffusion process driven by heterogeneity among firms in terms of returns to adoption, rather than in terms of access to information. Recall that the model developed in Chapter 4 builds upon this assumption. One could criticize these interpretations for the fact that they assume a diffusion process among a population of fictitious firms, without an explicit formulation of heterogeneity, and for the fact that they express diffusion in terms of energy intensity improvements rather than number of adopters. These objections can be abandoned by applying a third interpretation in which it is assumed that each sector adopts a new retrofit technology instantaneously, and subsequently explores its potential energy-saving potential only gradually through a learning-by-using process. In sum, the fact that NEMO does not explicitly include a representation of heterogeneous firms within a sector, prevents us from distinguishing between the different interpretations. This provides an extra argument for examining the impact of different curves to describe the adjustment process for retrofit technologies.

Mechanism

In addition to the concave curve describing the adjustment process for retrofit technologies and learning-by-using for replacement technologies, we add the following two S-shaped curves in order to analyse the sensitivity of NEMO for the form of these curves:

$$F^{ar}_{\tau,t} = F^{ir}_{\tau,t} \cdot \left[1 + \left(\frac{F^{ir}_{\tau,t} - F^{ar}_{\tau,t-1}}{F^{ar}_{\tau,t-1}} \right) e^{-\phi} \right]^{-1} \tag{7.3b}$$

$$F^{ar}_{\tau,t} = F^{ir}_{\tau,t} \cdot \left[1 + \left(\frac{F^{ir}_{\tau,t} - F^{ar}_{\tau,t-1}}{F^{ar}_{\tau,t-1}} \right) e^{-\phi(F^{ir}_{\tau,t} - F^{ar}_{\tau,t-1})} \right]^{-1} \tag{7.3c}$$

both with $\phi = \psi\left(\frac{I_t^{tot}}{K_t}\right)^{\varepsilon}$ and under the conditions that $t > \tau$ and $F_{\tau,t}^{ir} \leq F_{\tau,t-1}^{ar}$.

Equation (7.3b) yields the classic symmetric S-curve. Equation (7.3c) yields an asymmetric S-shaped curve since the parameter ϕ is no longer constant, but a function of the decreasing learning or diffusion potential. This formulation captures the available empirical evidence that diffusion slows down over time (see for example Geroski 2000, Dixon 1980). Potential explanations for this slowdown are to be found in saturation, decrease of 'evangelistic enthusiasm', increasing resistance of late adopters or the fact that late adopters do have characteristics that form a barrier for adoption anyway (uncertainty, lack of financial capabilities, complementarities, and so on). The different curves are illustrated in Figure 7.8.

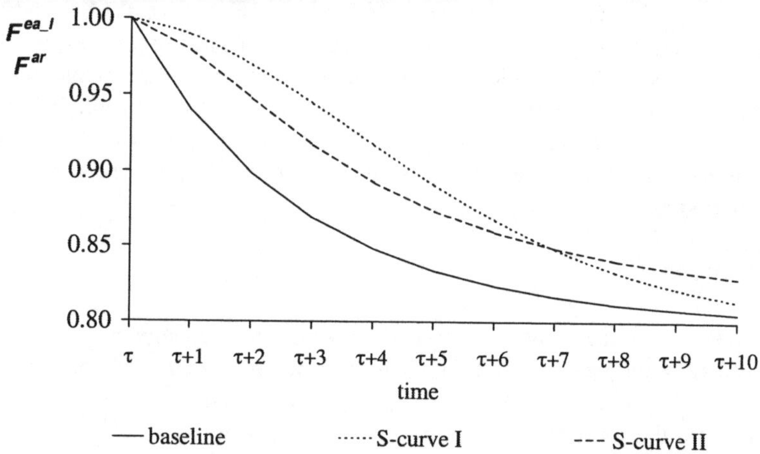

Figure 7.8 Energy intensity development for vintage τ during its lifetime under different forms of the adjustment process

Note that the curves are in reverse of the usual S-shaped diffusion curves because they are defined in terms of energy intensity rather than number of adopters. To illustrate the impact of the shape of the adjustment curve for retrofit technologies on the sequential steps of the model in determining the energy intensity level of the capital stock, Figure 7.9 again shows for the paper, pulp and printing sector the *ex ante* and *ex post* relative energy intensity level when the adjustment process takes the shape of the classic symmetric S-curve.

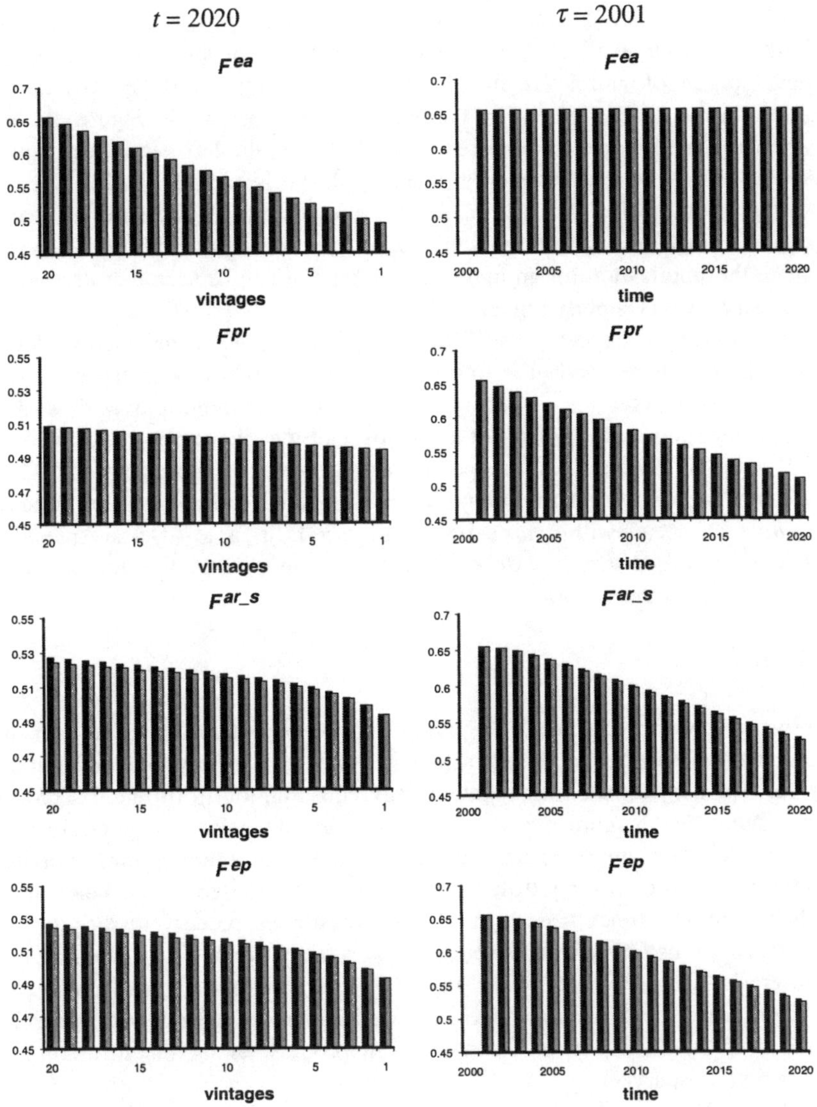

Figure 7.9 *Ex ante and ex post energy intensity in the Paper sector under Retrofit S-curve; (left) at time t=2020 with cross-section of different vintages in use and (right) for vintage τ=2001 during its lifetime. Baseline in grey.*

The left part of the figure again shows the relative energy intensity at time t = 2020 with a cross-section view of the different vintages in use with the most recent vintage at the right, and the right part of the figure shows the development of the relative energy intensity of a vintage installed in year τ = 2001 during its lifetime.[11] The baseline figures as shown in Figure 7.2 are again included in grey. Compared to the baseline, the left part of the figure shows that the *ex post* energy intensity level is slightly higher for all vintages, due to the fact that an S-curve implies an initially slow diffusion rate as compared to a concave curvature. The right part of the figure shows again the implication for an individual vintage: it can be seen that its energy intensity level is slightly higher.

Of course one expects these differences to vanish over time when the level of saturation is reached. It is to be noted however that this saturation level, defined by the ideal *ex post* energy intensity level, is moving over time, due to the emergence of new retrofit technologies. Since we apply a time-span of 20 years, it depends on the parameterization of the speed of diffusion whether the *ex post* energy intensity gap between a concave and an S-curve *ex post* disappears within this certain time period. In Table 7.B.4 in Appendix 7.B one can see that in case of a fast diffusion process the gap starts to decrease for some sectors.

Results

What are the consequences of applying an S-curve to describe the adjustment process for retrofit technologies together with the inclusion of a learning effect for the estimated aggregate energy consumption in different sectors? We found that the choice of the functional form does not change the impact of learning-by-using on energy demand,[12] while it has a limited impact on the adjustment process for retrofit technologies. This is due to the fact that a change in the functional form of the adjustment process applies to all vintages in use rather than to the newest only. Figure 7.10 shows the difference in use of electricity (Figure 7.10a) and fuel (Figure 7.10b) per sector over time as compared to the baseline, under constant energy prices and investment, and with the adjustment process taking the form of the symmetric S-curve.

It can be seen that the symmetric S-curve implies only an increase in energy use between 0.00-0.25 per cent over a 20-year period, as compared to the modified exponential function. This difference stems from the fact that in the S-shaped curve case, initial diffusion or learning is slow as compared to the case with a modified exponential function.

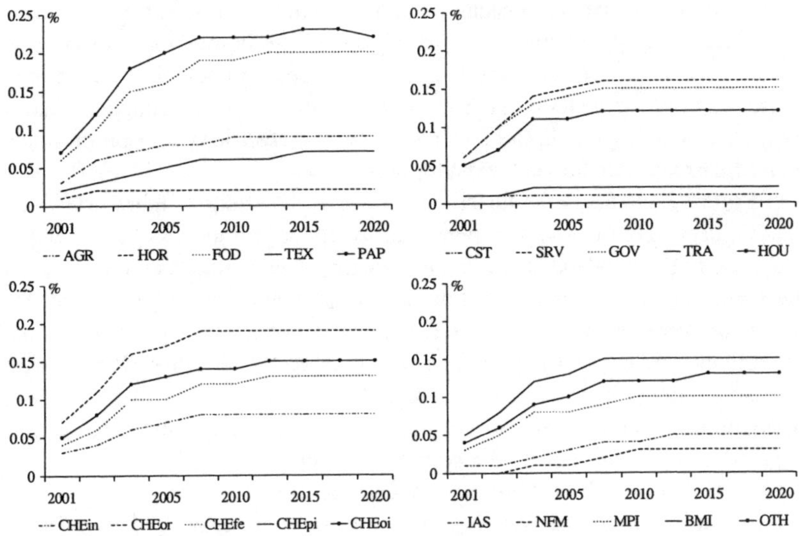

Figure 7.10a *Ceteris paribus effects of a symmetric S-shaped curve for retrofit adjustment on electricity use (under constant energy prices and investment); % change compared to baseline*

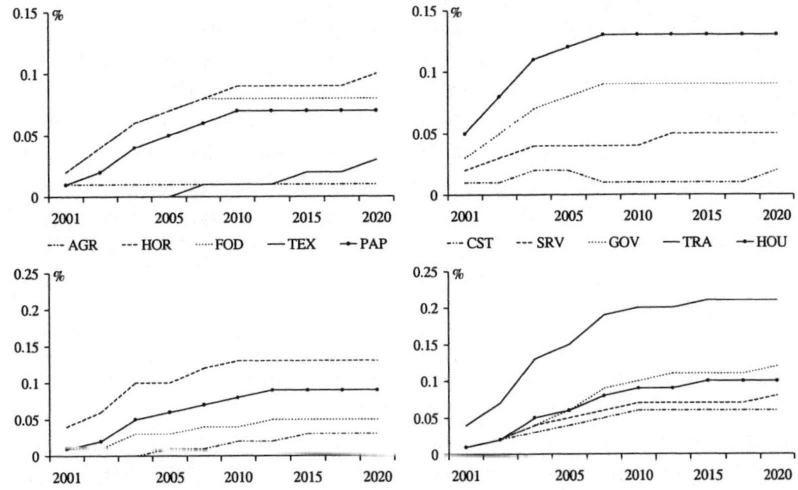

Figure 7.10b *Ceteris paribus effects of a symmetric S-shaped curve for retrofit adjustment on fuel use (under constant energy prices and investment); % change compared to baseline*

The differences between sectors can be explained from differences in the retrofit technology parameter γ. The higher γ, the higher the gap between potential and actual retrofit, and the higher the delay in actual *ex post* energy efficiency improvements in the S-curve case. As a result, the impact of an S-shaped curve is highest in the paper industry for electricity consumption and in building materials for fuel consumption.

Figure 7.11 shows the difference in use of electricity (Figure 7.11a) and fuel (Figure 7.11b) per sector over time for the asymmetric S-curve case as compared to the baseline, again under constant energy prices and investment. The figure again shows a very limited increase in electricity and fuel use, except for some sectors. The overall range lies between, respectively, 0 and 2.5 per cent, and 0 and 1 per cent, as compared to the baseline. Of course the increase in energy use is due to the inherent slowing down of diffusion under an asymmetric S-curve, while the differences between sectors reflect again the differences in the retrofit technology parameter γ.

Table 7.7 shows the difference in electricity and fuel use per sector over time as compared to the baseline, for both the symmetric and asymmetric curve, under the baseline fluctuations in energy prices and investment rates. These results confirm that in general the differences in electricity and fuel use are small, and highest in the case of an asymmetric curve. Moreover for most sectors the difference is higher for electricity as compared to fuel, due to the fact that $\gamma_e > \gamma_f$ for most sectors.

To single out the impact of energy policies, we ran the model including a permanent increase of the energy price of 5 per cent from 2005. The results are presented in Table 7.8, and show that under an S-shaped curve for retrofit investment, the impact of energy policies is higher in the short run, while the difference in impact disappears in the long run, due to the fact that the gap between the S-curve and the baseline modified exponential curve also vanishes. Moreover the difference in impact is higher under a symmetric S-curve, due to the larger gap between this curve and the baseline curve.

Sensitivity Analysis

Which value to choose for the parameter determining the diffusion speed? Decanio and Laitner (1997) present the results of a meta-analysis on estimates of the diffusion parameter for 298 different technologies taken from 17 studies. They found an average of the estimates of the diffusion rate parameter of around 0.4. So far we assumed the diffusion rate parameter to be 0.3. Therefore, we performed a sensitivity analysis, defining fast and slow diffusion by assuming a value of, respectively, 0.6 and 0.2 for the diffusion parameter. The sensitivity analysis revealed that in particular the asymmetric S-curve is rather sensitive to the speed of diffusion.[13]

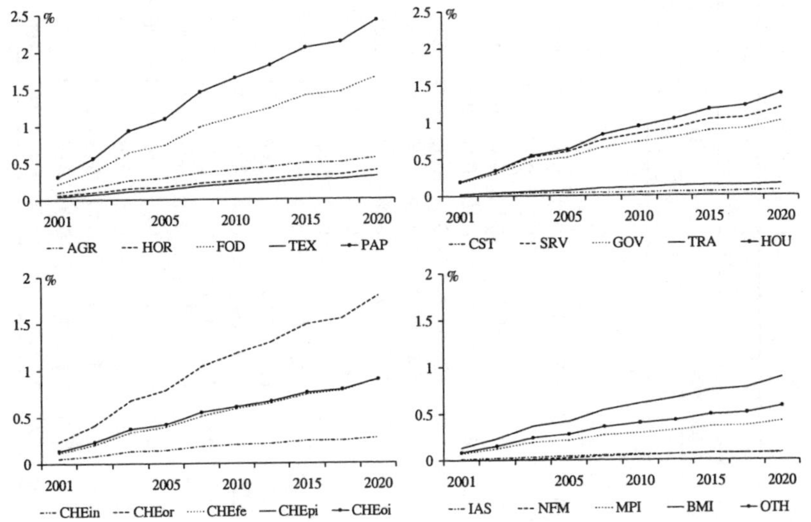

Fig. 7.11a Ceteris paribus effects of an asymmetric S-shaped curve for retrofit adjustment on electricity use (under constant energy prices and investment); % change compared to baseline

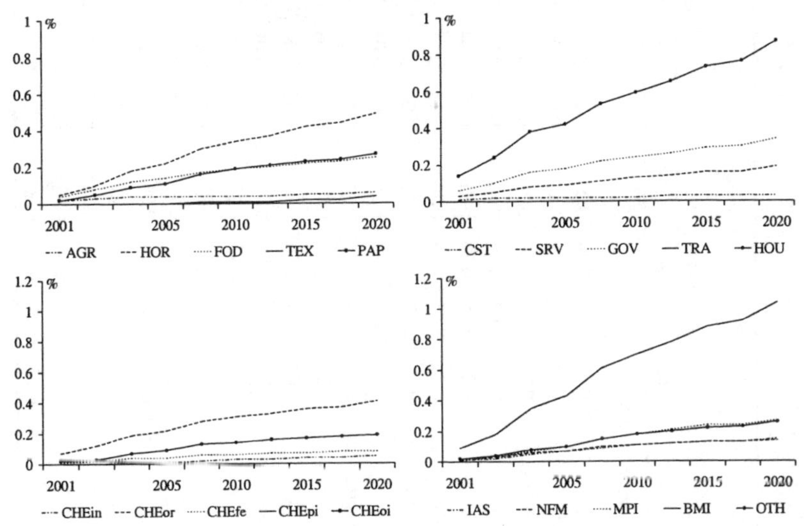

Fig. 7.11b Ceteris paribus effects of an asymmetric S-shaped curve for retrofit adjustment on fuel use (under constant energy prices and investment); % change compared to baseline

Table 7.7 Effects of a 5% increase in the technology parameters α and γ on energy use

	Symmetric S-curve								Asymmetric S-curve							
	Electricity				Fuel				Electricity				Fuel			
	2005	2010	2015	2020	2005	2010	2015	2020	2005	2010	2015	2020	2005	2010	2015	2020
% change compared to baseline																
Horticulture	0.08	0.06	0.04	0.04	0.10	0.06	0.10	0.12	0.54	0.57	0.54	0.53	0.31	0.21	0.39	0.52
Other Agriculture	0.18	0.17	0.16	0.15	0.05	0.02	0.02	0.03	0.58	0.66	0.72	0.79	0.11	0.06	0.05	0.09
Food & beverage industry	0.24	0.27	0.28	0.28	0.12	0.10	0.08	0.08	1.08	1.50	1.77	1.99	0.34	0.27	0.20	0.18
Textile industry	0.08	0.09	0.09	0.09	0.08	0.07	0.06	0.05	0.21	0.28	0.32	0.38	0.17	0.13	0.11	0.09
Paper industry	0.33	0.35	0.33	0.31	0.07	0.04	0.04	0.07	1.56	2.09	2.47	2.96	0.19	0.11	0.13	0.23
Organic chemicals	0.32	0.33	0.33	0.33	0.18	0.19	0.19	0.20	1.18	1.48	1.73	1.99	0.34	0.39	0.44	0.49
Inorganic chemicals	0.13	0.13	0.13	0.13	0.07	0.05	0.03	0.03	0.24	0.28	0.30	0.33	0.09	0.06	0.04	0.04
Fertilizers	0.19	0.20	0.20	0.20	0.04	0.03	0.03	0.05	0.58	0.74	0.87	1.00	0.06	0.04	0.04	0.07
Chemical products industry	0.24	0.25	0.25	0.24	0.07	0.05	0.05	0.08	0.67	0.82	0.93	1.05	0.13	0.08	0.08	0.14
Other chemical industry	0.24	0.25	0.25	0.24	0.08	0.05	0.05	0.08	0.67	0.82	0.93	1.05	0.13	0.08	0.08	0.14
Iron and steel	0.05	0.06	0.06	0.05	0.02	0.04	0.06	0.07	0.06	0.08	0.08	0.09	0.03	0.07	0.12	0.14
Non-ferrous metals	0.02	0.04	0.04	0.04	0.07	0.04	0.07	0.08	0.04	0.07	0.08	0.09	0.11	0.07	0.12	0.15
Metal products	0.16	0.16	0.15	0.15	0.09	0.05	0.07	0.09	0.36	0.43	0.49	0.56	0.18	0.10	0.14	0.23
Building materials	0.22	0.24	0.23	0.22	0.18	0.08	0.16	0.20	0.65	0.86	1.01	1.17	0.63	0.28	0.63	0.99
Other industry	0.16	0.18	0.18	0.17	0.10	0.05	0.08	0.09	0.42	0.55	0.64	0.75	0.20	0.11	0.19	0.26
Construction	0.10	0.06	0.04	0.04	0.03	0.03	0.03	0.03	0.24	0.18	0.14	0.12	0.05	0.05	0.05	0.06
Services	0.31	0.30	0.28	0.27	0.07	0.03	0.05	0.06	1.18	1.39	1.51	1.66	0.16	0.08	0.13	0.19
Government	0.31	0.29	0.26	0.25	0.12	0.08	0.11	0.12	1.15	1.30	1.37	1.49	0.27	0.18	0.28	0.36
Transport	0.05	0.05	0.03	0.03	0.00	0.00	0.00	0.00	0.18	0.21	0.19	0.19	0.00	0.00	0.00	0.00
Households	0.23	0.22	0.21	0.20	0.18	0.13	0.17	0.19	1.15	1.39	1.53	1.69	0.65	0.47	0.67	0.86

Table 7.8 Effects of a 5% increase in energy price in 2005 on energy use, under S-curves for retrofit

	Symmetric S-curve								Asymmetric S-curve							
	Electricity				Fuel				Electricity				Fuel			
	2005	2010	2015	2020	2005	2010	2015	2020	2005	2010	2015	2020	2005	2010	2015	2020
% change compared to baseline																
Horticulture	-5.26	-2.94	-2.50	0.00	-5.56	-2.13	0.00	0.00	-21.05	-17.65	-10.00	-4.44	-11.11	-6.38	-1.56	-1.23
Other Agriculture	-4.00	-1.52	0.00	0.00	-6.90	-2.70	-1.05	-0.92	-8.00	-4.55	-2.08	-0.85	-17.24	-5.41	-1.05	-0.92
Food & beverage industry	-2.70	-1.72	0.00	0.00	-3.45	-2.38	0.00	-0.81	-8.11	-10.34	-4.84	-3.13	-10.34	-7.14	-2.83	-1.63
Textile industry	0.00	-3.23	0.00	0.00	-2.17	-1.32	-0.98	0.00	-5.88	-6.45	-2.78	0.00	-2.17	-1.32	-1.96	-0.86
Paper industry	0.00	0.00	0.00	0.00	0.00	0.00	0.00	0.00	-10.53	-11.11	-7.14	-4.35	-4.17	-2.50	-0.81	-0.62
Organic chemicals	-4.35	-2.27	0.00	0.00	0.00	0.00	0.00	0.00	-13.04	-11.36	-6.25	-4.08	-5.26	-2.86	-2.44	-2.22
Inorganic chemicals	-5.26	-2.86	0.00	0.00	-5.88	-4.88	-2.17	0.00	-5.26	-2.86	-2.44	-2.22	-5.88	-4.88	-2.17	-2.13
Fertilizers	-5.00	-2.70	0.00	0.00	0.00	0.00	0.00	0.00	-10.00	-8.11	-2.38	-2.17	-6.67	-3.23	-2.70	0.00
Chemical products industry	-4.00	-1.82	-1.52	0.00	-8.70	-1.92	-1.67	0.00	-12.00	-9.09	-4.55	-1.37	-8.70	-3.85	-1.67	-1.56
Other chemical industry	-4.00	-1.82	-1.52	0.00	-8.70	-1.92	-1.67	0.00	-12.00	-9.09	-4.55	-1.37	-8.70	-3.85	-1.67	-1.56
Iron and steel	0.00	0.00	-2.44	0.00	0.00	-2.50	0.00	-1.54	0.00	-2.94	-2.44	-2.22	-6.25	-2.50	-1.85	-1.54
Non-ferrous metals	-5.56	-2.94	0.00	0.00	-6.67	0.00	0.00	0.00	-5.56	-5.88	0.00	0.00	-6.67	-3.33	-2.70	-2.44
Metal products	-4.76	-2.50	-2.22	0.00	-6.45	-1.59	-1.39	0.00	-9.52	-7.50	-4.44	-2.17	-9.68	-4.76	-2.78	-1.33
Building materials	-4.76	-2.50	0.00	0.00	-5.41	-1.98	0.00	0.00	-9.52	-10.00	-2.27	-2.13	-16.22	-8.91	-4.03	-2.84
Other industry	0.00	-3.03	0.00	0.00	-4.76	-2.00	0.00	0.00	-5.56	-6.06	-2.63	-2.44	-9.52	-6.00	-3.45	-1.61
Construction	-4.17	-2.22	-2.04	0.00	0.00	0.00	-1.92	0.00	-12.50	-6.67	-4.08	0.00	-5.88	-2.50	-1.92	0.00
Services	-2.33	-1.27	0.00	0.00	-8.70	-1.79	-1.49	0.00	-9.30	-7.59	-3.26	-1.94	-13.04	-7.14	-4.48	-1.39
Government	-2.44	-2.17	-0.87	0.00	-4.35	-3.39	0.00	-1.30	-12.20	-8.70	-4.35	-2.24	-13.04	-8.47	-2.86	-2.60
Transport	-3.85	0.00	0.00	0.00	0.00	0.00	0.00	0.00	-15.38	-6.25	-1.69	0.00	0.00	0.00	0.00	0.00
Households	-2.94	-1.25	0.00	0.00	-1.08	-1.41	-0.63	0.00	-11.76	-7.50	-2.61	-2.07	-5.38	-6.34	-4.40	-2.98

7.5 CONCLUSIONS

Understanding the interaction between technological change, energy use and environmental policy is a prerequisite for designing effective policies as well as for properly forecasting their impact. In this chapter we have examined the effect of microeconomic characteristics of technology adoption and diffusion on sectoral energy consumption in NEMO, an applied economic model of economy–energy interaction used for policy analysis in the Netherlands. Building upon insights developed in the previous chapters, we introduced mechanisms into the model representing technological complementarity and a learning-by-using process for new energy-saving technologies. Moreover, we refined the external learning or diffusion process by which technologies are upgraded. We analysed to what extent the results of an applied economic policy model such as NEMO are sensitive to theoretical mechanisms such as those developed in the stylized theoretical models of Chapters 3 and 4.

We found the effect of including technological complementarity to be rather small. This result is at least partly due to the restrictive way in which it has been modelled – caused by the impossibility of deriving a demand curve for specific vintages in NEMO. To fully assess its role in determining aggregate energy demand would require a complete restructuring of NEMO, which comes close to a calibration of the model developed in Chapter 3, using the database underlying NEMO.

In contrast our analysis showed that introducing learning-by-using for new energy-saving technologies in NEMO led to a considerable increase in sectoral energy demand, due to the assumption that the productivity gains inherent in the adoption of new technologies are only realized after some period of time, while the energy demand forecasts are also rather sensitive to the assumed speed of learning. This result is in line with the conclusion drawn from the theoretical analysis in Chapter 3, suggesting that learning-by-using processes can be an important part of the answer to the question why a substantial unexplored energy-saving potential exists, according to bottom-up projections such as ICARUS. Moreover it suggests that policies aiming to achieve energy efficiency improvements should pay deliberate attention to learning processes and barriers to technology adoption raised by lack of information or expected loss of expertise.

In sum, the findings in this chapter underline the relevance of including technology diffusion in applied economic policy models. Until now, most macroeconomic models including endogenous technological change have been characterized by much emphasis on innovation through R&D and learning-by-doing processes, the latter usually being a costless by-product of cumulative investment, installed capacity or abatement. According to these models, aggregate knowledge accumulation will enhance energy efficiency improvements (and thus lower emissions) through technological change, while lowering its costs. Our analysis however showed that the inclusion of a more detailed description of technology diffusion processes in a macroeconomic model might yield an increase in energy demand – at least in the short run – due to discontinuities and temporary loss of expertise inherent in processes of technological change.

APPENDIX 7.A

Table 7.A.1 Sector classification

	Sector	Code
1	Horticulture	HOR
2	Other Agriculture	AGR
3	Food & beverage industry	FOD
4	Textile industry	TEX
5	Paper industry	PAP
6	Organic chemicals	CHEor
7	Inorganic chemicals	CHEin
8	Fertilizers	CHEfe
9	Chemical products industry	CHEpi
10	Other chemical industry	CHEoi
11	Iron and steel	IAS
12	Non-ferrous metals	NFM
13	Metal products	MPI
14	Building materials	BMI
15	Other industry	OTH
16	Construction	CST
17	Services	SRV
18	Government	GOV
19	Transport	TRA
20	Households	HOU

Table 7.A.2 Elasticities, trends and scrapping ages

	Trends				Elasticities						Scrapping Ages	
	Electricity		Fuel		Electricity			Fuel				
	Repl.	Retrof.	Repl.	Retrof.	Repl.	Retrof.	Good-	Repl.	Retrof.	Good-	a_{min}	a_{max}
	α_e	γ_e	α_f	γf	β_e	δ_e	θ_e	β_f	δ_f	θ_f		
Horticulture	4.20	0.10	1.20	0.38	0.08	0.04	0.02	0.18	0.04	0.02	15	25
Other Agriculture	1.24	0.50	0.89	0.10	0.23	0.04	0.02	0.22	0.10	0.02	10	20
Food & beverage industry	1.14	1.02	0.38	0.34	0.08	0.07	0.06	0.33	0.14	0.02	25	35
Textile industry	1.01	0.24	0.16	0.10	0.08	0.04	0.02	0.18	0.17	0.08	30	40
Paper industry	1.50	1.34	0.90	0.25	0.08	0.05	0.02	0.45	0.06	0.02	18	28
Organic chemicals	1.30	1.17	0.38	0.64	0.08	0.07	0.02	0.08	0.04	0.02	18	28
Inorganic chemicals	0.48	0.40	0.16	0.10	0.08	0.04	0.02	0.08	0.08	0.02	18	28
Fertilizers	1.28	0.63	0.19	0.17	0.08	0.05	0.02	0.08	0.04	0.02	18	28
Chemical products industry	0.90	0.81	0.33	0.30	0.15	0.07	0.02	0.13	0.09	0.02	18	28
Other chemical industry	0.90	0.81	0.33	0.30	0.15	0.07	0.02	0.13	0.09	0.02	18	28
Iron and steel	0.21	0.16	0.54	0.20	0.08	0.04	0.02	0.15	0.04	0.02	18	28
Non-ferrous metals	0.57	0.10	0.29	0.26	0.08	0.04	0.02	0.08	0.04	0.02	18	28
Metal products	0.53	0.47	0.39	0.35	0.08	0.06	0.02	0.14	0.10	0.03	30	40
Building materials	0.86	0.80	0.96	0.84	0.08	0.06	0.02	0.33	0.16	0.02	18	28
Other industry	0.64	0.58	0.42	0.32	0.08	0.04	0.02	0.14	0.09	0.02	30	40
Construction	0.85	0.10	0.16	0.10	0.08	0.07	0.02	0.13	0.04	0.02	14	24
Services	1.07	0.96	0.73	0.20	0.18	0.09	0.06	0.18	0.10	0.02	15[1]	25[1]
Government	0.97	0.87	0.46	0.41	0.27	0.11	0.04	0.19	0.10	0.02	15[1]	25[1]
Transport	2.10	0.10	0.75	0.00	0.10	0.05	0.03	0.40	0.00	0.05	10	15
Households	2.26	0.63	0.90	0.63	0.31	0.05	0.04	0.28	0.15	0.15	10[2]	25[2]

Notes: [1] Electricity value. Fuel values for a_{min} and a_{max} are, respectively, 37 and 47. [2] Fuel values for a_{min} and a_{max} are, respectively, 40 and 50

Table 7.A.3 Marginal investment growth in variant 2002 in %

	2001	2002	2003	2004	2005	2006	2007	2008	2009	2010	2011	...	2020
Horticulture	-18.0	14.5	0.0	0.0	0.0	0.0	0.3	0.3	0.3	0.3	-0.1	...	-0.1
Other Agriculture	-18.0	14.5	0.0	0.0	0.0	0.0	0.3	0.3	0.3	0.3	-0.1	...	-0.1
Food & beverage industry	-15.4	0.2	0.9	0.9	0.9	0.9	0.7	0.7	0.7	0.7	0.7	...	0.7
Textile industry	1.4	3.5	-0.7	-0.7	-0.7	-0.7	2.4	2.4	2.4	2.4	-4.8	...	-4.8
Paper industry	1.4	3.5	-0.7	-0.7	-0.7	-0.7	2.4	2.4	2.4	2.4	-4.8	...	-4.8
Organic chemicals	5.6	5.3	2.8	2.8	2.8	2.9	2.1	2.1	2.1	2.1	2.1	...	2.1
Inorganic chemicals	5.6	5.3	2.8	2.8	2.8	2.9	2.1	2.1	2.1	2.1	2.1	...	2.1
Fertilizers	5.6	5.3	2.8	2.8	2.8	2.8	2.1	2.1	2.1	2.1	2.1	...	2.1
Chemical products industry	5.6	5.3	2.8	2.8	2.8	2.9	2.1	2.1	2.1	2.1	2.1	...	2.1
Other chemical industry	5.6	5.4	2.9	2.9	2.9	2.9	2.1	2.1	2.1	2.1	2.1	...	2.1
Iron and steel	3.9	3.6	-0.9	-0.9	-0.9	-0.9	-2.1	-2.1	-2.1	-2.1	-4.5	...	-4.5
Non-ferrous metals	3.9	3.6	-0.9	-0.9	-0.9	-0.9	-2.1	-2.1	-2.1	-2.1	-4.5	...	-4.5
Metal products	3.9	3.7	-0.9	-0.9	-0.9	-0.9	-2.1	-2.1	-2.1	-2.1	-4.5	...	-4.5
Building materials	1.4	0.2	-5.4	-5.4	-5.4	-5.4	2.4	2.4	2.4	2.4	-4.8	...	-4.8
Other industry	1.4	0.2	-5.4	-5.4	-5.4	-5.4	2.4	2.4	2.4	2.4	-4.8	...	-4.8
Construction	2.3	1.2	1.0	1.0	1.0	1.0	1.7	1.7	1.7	1.7	1.3	...	1.3
Services	0.2	-2.2	-2.0	-2.0	-2.0	-2.0	1.5	1.5	1.5	1.5	-1.8	...	-1.8
Government	5.1	-3.7	-0.2	-0.2	-0.2	-0.2	0.8	0.8	0.8	0.8	-1.1	...	-1.1
Transport	0.0	-0.7	-2.4	-2.4	-2.4	-2.4	2.2	2.2	2.2	2.2	-1.8	...	-1.8
Households	1.1	2.1	2.4	2.4	2.4	2.4	2.5	2.5	2.5	2.5	2.0	...	2.0

Table 7.A.4 Marginal changes in the price of electricity in variant 2002 in %

	2001	2002	2003	2004	2005	2006	2007	2008	2009	2010	2011 …	2020
Horticulture	16.0	-2.5	0.8	1.9	1.0	0.7	0.7	0.7	1.0	1.1	0.0 …	0.0
Other Agriculture	16.0	-2.5	0.8	1.9	1.0	0.7	0.7	0.7	1.0	1.1	0.0 …	0.0
Food & beverage industry	11.7	-3.0	0.6	0.3	0.3	0.3	0.3	0.3	0.9	1.1	0.0 …	0.0
Textile industry	11.7	-3.0	0.6	0.3	0.3	0.3	0.3	0.3	0.9	1.1	0.0 …	0.0
Paper industry	11.7	-3.0	0.6	0.3	0.3	0.3	0.3	0.3	0.9	1.1	0.0 …	0.0
Organic chemicals	11.7	-3.0	0.6	0.3	0.3	0.3	0.3	0.3	0.9	1.1	0.0 …	0.0
Inorganic chemicals	11.7	-3.0	0.6	0.3	0.3	0.3	0.3	0.3	0.9	1.1	0.0 …	0.0
Fertilizers	11.7	-3.0	0.6	0.3	0.3	0.3	0.3	0.3	0.9	1.1	0.0 …	0.0
Chemical products industry	11.7	-3.0	0.6	0.3	0.3	0.3	0.3	0.3	0.9	1.1	0.0 …	0.0
Other chemical industry	11.7	-3.0	0.6	0.3	0.3	0.3	0.3	0.3	0.9	1.1	0.0 …	0.0
Iron and steel	11.7	-3.0	0.6	0.3	0.3	0.3	0.3	0.3	0.9	1.1	0.0 …	0.0
Non-ferrous metals	11.7	-3.0	0.6	0.3	0.3	0.3	0.3	0.3	0.9	1.1	0.0 …	0.0
Metal products	11.7	-3.0	0.6	0.3	0.3	0.3	0.3	0.3	0.9	1.1	0.0 …	0.0
Building materials	11.7	-3.0	0.6	0.3	0.3	0.3	0.3	0.3	0.9	1.1	0.0 …	0.0
Other industry	11.7	-3.0	0.6	0.3	0.3	0.3	0.3	0.3	0.9	1.1	0.0 …	0.0
Construction	16.0	-2.5	0.8	1.9	1.0	0.7	0.7	0.7	1.0	1.1	0.0 …	0.0
Services	16.0	-2.5	0.8	1.9	1.0	0.7	0.7	0.7	1.0	1.1	0.0 …	0.0
Government	16.0	-2.5	0.8	1.9	1.0	0.7	0.7	0.7	1.0	1.1	0.0 …	0.0
Transport	11.7	-3.0	0.8	1.9	1.0	0.7	0.7	0.7	1.0	1.1	0.0 …	0.0
Households	16.0	-2.5	0.8	1.9	1.0	0.7	0.7	0.7	1.0	1.1	0.0 …	0.0

Table 7.A.5 Marginal changes in the price of fuel in variant 2002 in %

	2001	2002	2003	2004	2005	2006	2007	2008	2009	2010	2011	...	2020
Horticulture	15.5	-19.2	-0.1	-5.8	-6.2	-7.4	-8.0	-10.6	-1.3	1.3	0.0	...	0.0
Other Agriculture	9.2	-13.2	0.4	-2.6	-2.9	-3.6	-3.6	-4.4	-0.4	0.6	0.0	...	0.0
Food & beverage industry	16.5	-19.4	-0.3	-6.0	-6.3	-7.5	-8.1	-10.7	-1.3	1.3	0.0	...	0.0
Textile industry	17.1	-19.6	0.0	-6.1	-6.5	-7.7	-8.5	-11.2	-1.4	1.3	0.0	...	0.0
Paper industry	17.0	-19.6	0.0	-6.1	-6.5	-7.7	-8.5	-11.2	-1.4	1.3	0.0	...	0.0
Organic chemicals	-8.2	-13.6	-1.5	-0.7	-1.0	-2.9	-0.4	-1.0	0.8	1.2	0.0	...	0.0
Inorganic chemicals	-1.5	-14.7	-2.2	-1.9	-2.0	-3.6	-1.6	-2.4	0.5	1.2	0.0	...	0.0
Fertilizers	17.2	-19.6	0.0	-6.1	-6.6	-7.8	-8.5	-11.2	-1.4	1.3	0.0	...	0.0
Chemical products industry	15.3	-19.2	-0.1	-5.7	-6.1	-7.4	-7.9	-10.5	-1.2	1.3	0.0	...	0.0
Other chemical industry	16.7	-19.4	-0.3	-6.0	-6.3	-7.5	-8.1	-10.7	-1.3	1.3	0.0	...	0.0
Iron and steel	12.6	-15.5	-7.5	-3.2	-2.5	-3.4	-1.0	-0.5	0.8	1.2	0.0	...	0.0
Non-ferrous metals	6.1	-17.0	-0.6	-3.8	-4.1	-5.6	-5.0	-6.8	-0.5	1.3	0.0	...	0.0
Metal products	7.0	-17.2	-0.6	-4.0	-4.3	-5.8	-5.3	-7.1	-0.5	1.3	0.0	...	0.0
Building materials	14.0	-18.6	-0.7	-5.3	-5.6	-6.9	-7.0	-9.3	-1.0	1.3	0.0	...	0.0
Other industry	1.1	-15.8	-0.9	-2.7	-3.0	-4.7	-3.4	-4.8	0.0	1.2	0.0	...	0.0
Construction	-6.2	-12.9	-3.1	-0.5	-0.5	-2.1	0.7	0.8	1.1	1.1	0.0	...	0.0
Services	7.2	-13.2	0.2	-2.4	-2.6	-3.4	-3.2	-3.9	-0.2	0.7	0.0	...	0.0
Government	9.4	-13.3	0.4	-2.7	-2.9	-3.6	-3.6	-4.5	-0.4	0.6	0.0	...	0.0
Transport	-14.2	-12.2	-1.8	0.5	0.3	-1.7	1.5	1.4	1.3	1.2	0.0	...	0.0
Households	10.3	-13.3	0.5	-2.8	-3.1	-3.7	-3.8	-4.7	-0.5	0.6	0.0	...	0.0

Table 7.A.6 1999 electricity and fuel price levels as compared to base year t_0 (t_0=1)

	1999	1999
	P_{elec}	P_{fuel}
Horticulture	1.14	0.95
Other Agriculture	1.14	1.17
Food & beverage industry	1.09	1.05
Textile industry	1.03	0.94
Paper industry	1.03	0.94
Organic chemicals	1.03	0.92
Inorganic chemicals	1.03	0.92
Fertilizers	0.88	0.94
Chemical products industry	1.03	0.92
Other chemical industry	1.03	0.92
Iron and steel	0.73	0.74
Non-ferrous metals	0.73	0.74
Metal products	1.23	0.82
Building materials	1.09	0.89
Other industry	1.03	0.92
Construction	1.14	1.19
Services	1.15	1.05
Government	1.15	1.05
Transport	1.06	1.07
Households	1.15	1.10

Table 7.B.1 Effects of a high and a low complementarity effect on energy use

	Electricity								Fuel							
	High complementarity				Low complementarity				High complementarity				Low complementarity			
	2005	2010	2015	2020	2005	2010	2015	2020	2005	2010	2015	2020	2005	2010	2015	2020
% change compared to baseline																
Horticulture	0.00	0.00	-0.01	-0.03	0.00	0.00	0.00	0.00	0.02	-0.01	0.02	-0.01	0.00	0.00	0.00	0.00
Other Agriculture	0.08	0.08	0.03	0.04	0.01	0.01	0.00	0.00	0.00	0.03	0.04	0.01	0.00	0.00	0.01	0.00
Food & beverage industry	-0.07	-0.08	-0.2	-0.36	-0.01	-0.01	-0.03	-0.05	-0.08	0.12	-0.09	-0.18	-0.01	0.02	0.02	-0.01
Textile industry	-0.19	-0.05	0.02	-0.01	-0.02	-0.01	0.00	0.00	0.00	0.10	0.15	0.19	0.00	0.01	0.02	0.02
Paper industry	0.05	-0.09	0.08	0.07	0.00	-0.02	0.00	0.00	-0.05	-0.09	-0.06	0.00	-0.01	0.01	0.00	0.00
Organic chemicals	0.25	0.39	0.58	0.47	0.03	0.04	0.07	0.05	0.13	0.21	0.31	0.25	0.01	0.02	0.03	0.03
Inorganic chemicals	0.08	0.12	0.19	0.15	0.01	0.01	0.02	0.02	0.01	0.02	0.04	0.05	0.00	0.01	0.01	0.01
Fertilizers	0.12	0.19	0.29	0.23	0.01	0.02	0.03	0.03	-0.01	0.01	0.04	0.06	0.00	0.01	0.01	0.01
Chemical products industry	0.17	0.26	0.39	0.31	0.02	0.03	0.04	0.03	-0.02	0.06	0.06	0.10	0.00	0.01	0.01	0.01
Other chemical industry	0.17	0.26	0.39	0.31	0.02	0.03	0.04	0.03	-0.01	0.07	0.06	0.11	0.00	0.02	0.01	0.01
Iron and steel	-0.01	0.01	0.08	0.05	0.00	0.00	0.01	0.00	-0.09	-0.03	0.09	0.06	-0.01	0.00	0.01	0.00
Non-ferrous metals	-0.04	-0.03	0.02	0.01	-0.01	0.00	0.00	0.00	-0.07	-0.07	0.08	0.05	-0.01	0.00	0.01	0.01
Metal products	0.03	0.11	0.09	0.01	0.00	0.01	0.01	0.00	-0.04	-0.10	-0.05	-0.04	0.00	0.00	0.00	0.00
Building materials	0.20	0.26	0.18	-0.02	0.02	0.03	0.02	0.00	0.05	-0.17	0.06	-0.04	0.01	0.00	0.01	0.00
Other industry	-0.01	0.14	0.19	0.13	0.00	0.01	0.02	0.01	-0.06	-0.06	0.05	0.05	-0.01	0.00	0.01	0.01
Construction	0.03	0.04	0.02	0.01	0.00	0.00	0.00	0.00	0.01	0.02	0.02	0.01	0.00	0.00	0.00	0.00
Services	0.14	0.00	-0.07	-0.15	0.01	0.00	-0.01	-0.02	0.05	-0.03	-0.01	-0.04	0.01	0.00	0.00	0.00
Government	0.2	0.08	-0.02	-0.06	0.02	0.01	0.00	-0.01	0.01	-0.06	0.00	-0.01	0.00	0.00	0.00	0.00
Transport	0.02	0.01	0.01	0.01	0.00	0.00	0.00	0.00	0.00	0.00	0.00	0.00	0.00	0.00	0.00	0.00
Households	0.11	0.11	0.12	0.15	0.01	0.01	0.01	0.02	0.07	-0.15	-0.27	-0.40	0.01	-0.01	-0.03	-0.05

Table 7.B.2 Effects of a fast and slow speed of learning-by-using on energy use

	Electricity								Fuel							
	Fast learning				Slow learning				Fast learning				Slow learning			
	2005	2010	2015	2020	2005	2010	2015	2020	2005	2010	2015	2020	2005	2010	2015	2020
% change compared to baseline																
Horticulture	0.48	0.45	0.41	0.4	1.70	2.62	3.09	3.30	0.64	0.66	0.61	0.59	2.12	3.19	3.65	3.81
Other Agriculture	0.60	0.59	0.59	0.6	2.52	3.72	4.15	4.27	0.85	1.1	0.79	0.61	2.69	4.17	4.42	4.29
Food & beverage industry	0.62	0.67	0.68	0.69	1.40	2.3	2.83	3.16	0.82	1.39	1.12	0.86	1.45	2.79	3.33	3.49
Textile industry	0.63	0.72	0.66	0.57	1.29	2.23	2.63	2.73	1.05	1.97	1.86	1.67	1.55	3.20	3.93	4.09
Paper industry	0.72	0.73	0.65	0.55	2.00	3.04	3.52	3.63	0.76	1.07	0.79	0.63	1.90	3.23	3.73	3.77
Organic chemicals	0.73	0.68	0.63	0.59	2.53	3.44	3.68	3.66	0.76	0.71	0.66	0.61	2.61	3.52	3.74	3.71
Inorganic chemicals	0.73	0.67	0.62	0.58	2.55	3.44	3.67	3.63	0.92	0.87	0.71	0.61	2.68	3.64	3.78	3.69
Fertilizers	0.70	0.65	0.6	0.56	2.46	3.35	3.59	3.56	0.82	0.92	0.66	0.60	2.60	3.64	3.77	3.68
Chemical products industry	0.72	0.67	0.62	0.58	2.53	3.43	3.67	3.64	0.97	1.22	0.76	0.60	2.66	3.91	3.92	3.74
Other chemical industry	0.72	0.67	0.62	0.58	2.53	3.43	3.67	3.64	0.98	1.25	0.77	0.61	2.66	3.93	3.94	3.74
Iron and steel	0.75	0.72	0.61	0.50	2.17	3.23	3.58	3.52	0.79	0.73	0.61	0.49	2.18	3.24	3.58	3.51
Non-ferrous metals	0.72	0.69	0.59	0.48	2.10	3.15	3.51	3.46	0.77	0.74	0.63	0.52	2.18	3.25	3.61	3.55
Metal products	0.70	0.62	0.49	0.39	1.89	2.62	2.76	2.73	0.83	0.86	0.53	0.41	1.97	2.81	2.87	2.79
Building materials	0.65	0.58	0.49	0.41	2.16	2.97	3.27	3.39	0.82	0.88	0.57	0.46	2.21	3.22	3.44	3.52
Other industry	0.65	0.61	0.52	0.41	1.71	2.44	2.7	2.70	0.73	0.67	0.54	0.43	1.75	2.50	2.73	2.72
Construction	0.66	0.60	0.56	0.56	2.49	3.38	3.69	3.82	0.76	0.66	0.62	0.62	2.61	3.50	3.81	3.94
Services	0.64	0.62	0.59	0.57	2.12	3.11	3.64	3.85	0.65	0.64	0.48	0.39	1.61	2.29	2.45	2.42
Government	0.68	0.66	0.65	0.64	2.15	3.15	3.68	3.90	0.68	0.66	0.59	0.54	1.58	2.32	2.61	2.68
Transport	0.50	0.44	0.42	0.40	2.99	4.03	4.36	4.37	0.66	0.64	0.63	0.61	3.08	4.27	4.60	4.60
Households	0.67	0.72	0.75	0.76	2.13	3.38	3.91	4.08	0.66	0.69	0.68	0.67	1.38	2.17	2.55	2.75

Table 7.B.3 Effects of 5% and 20% learning potential on energy use, under learning-by-using

	Electricity								Fuel							
	Learning Potential 5%				Learning Potential 20%				Learning Potential 5%				Learning Potential 20%			
	2005	2010	2015	2020	2005	2010	2015	2020	2005	2010	2015	2020	2005	2010	2015	2020
% change compared to baseline																
Horticulture	0.68	0.88	0.92	0.93	2.70	3.50	3.66	3.69	0.86	1.12	1.16	1.16	3.40	4.42	4.60	4.58
Other Agriculture	0.99	1.23	1.29	1.31	3.90	4.88	5.10	5.19	1.15	1.71	1.53	1.33	4.16	5.48	5.40	5.21
Food & beverage industry	0.61	0.88	0.98	1.03	2.43	3.49	3.90	4.09	0.70	1.39	1.43	1.25	2.51	4.19	4.52	4.47
Textile industry	0.57	0.86	0.91	0.87	2.25	3.42	3.64	3.46	0.78	1.60	1.87	1.86	2.72	4.96	5.36	5.09
Paper industry	0.84	1.10	1.14	1.09	3.32	4.35	4.54	4.31	0.82	1.32	1.28	1.16	3.14	4.70	4.88	4.55
Organic chemicals	1.02	1.19	1.18	1.13	4.03	4.73	4.67	4.48	1.05	1.22	1.21	1.16	4.15	4.84	4.78	4.59
Inorganic chemicals	1.02	1.19	1.17	1.13	4.03	4.70	4.65	4.46	1.17	1.39	1.28	1.17	4.22	4.94	4.77	4.51
Fertilizers	0.98	1.16	1.14	1.09	3.90	4.57	4.52	4.33	1.07	1.40	1.24	1.15	4.11	4.98	4.77	4.52
Chemical products industry	1.02	1.19	1.18	1.13	4.02	4.70	4.65	4.47	1.16	1.71	1.37	1.18	4.20	5.33	4.91	4.56
Other chemical industry	1.02	1.19	1.18	1.13	4.02	4.70	4.65	4.47	1.16	1.74	1.39	1.18	4.21	5.36	4.93	4.56
Iron and steel	0.90	1.14	1.14	1.04	3.56	4.54	4.51	4.12	0.92	1.15	1.14	1.03	3.57	4.56	4.51	4.09
Non-ferrous metals	0.87	1.11	1.11	1.01	3.44	4.41	4.41	4.02	0.91	1.16	1.15	1.05	3.58	4.59	4.57	4.18
Metal products	0.79	0.95	0.90	0.82	3.15	3.77	3.58	3.27	0.88	1.15	0.98	0.86	3.26	4.02	3.70	3.34
Building materials	0.88	1.04	1.03	0.99	3.48	4.10	4.08	3.92	0.96	1.30	1.13	1.05	3.57	4.49	4.33	4.11
Other industry	0.72	0.90	0.90	0.83	2.87	3.56	3.56	3.29	0.77	0.95	0.92	0.84	2.93	3.64	3.60	3.32
Construction	0.99	1.15	1.16	1.16	3.91	4.53	4.57	4.61	1.07	1.21	1.21	1.22	4.09	4.73	4.78	4.82
Services	0.87	1.09	1.15	1.16	3.43	4.31	4.57	4.61	0.71	0.90	0.83	0.76	2.68	3.31	3.23	2.98
Government	0.88	1.11	1.18	1.20	3.49	4.38	4.67	4.75	0.69	0.88	0.90	0.87	2.69	3.47	3.56	3.44
Transport	1.11	1.26	1.31	1.30	4.39	4.96	5.17	5.12	1.18	1.39	1.43	1.42	4.64	5.47	5.66	5.62
Households	0.86	1.17	1.26	1.29	3.43	4.64	4.99	5.10	0.61	0.84	0.91	0.93	2.40	3.33	3.60	3.68

Table 7.B.4a *Effects of a fast and slow speed of adjustment for retrofit on energy use under symmetric S-curve*

	Electricity								Fuel							
	Fast adjustment				Slow adjustment				Fast adjustment				Slow adjustment			
	2005	2010	2015	2020	2005	2010	2015	2020	2005	2010	2015	2020	2005	2010	2015	2020
% change compared to baseline																
Horticulture	0.05	0.05	0.04	0.03	0.09	0.07	0.05	0.05	0.07	0.05	0.07	0.08	0.10	0.10	0.17	0.17
Other Agriculture	0.11	0.11	0.11	0.1	0.26	0.24	0.22	0.22	0.03	0.02	0.01	0.02	0.25	0.12	0.05	0.04
Food & beverage industry	0.09	0.11	0.11	0.12	0.49	0.47	0.45	0.45	0.07	0.05	0.04	0.05	0.26	0.21	0.19	0.13
Textile industry	0.05	0.06	0.07	0.06	0.14	0.13	0.11	0.11	0.04	0.03	0.03	0.02	0.42	0.33	0.27	0.22
Paper industry	0.14	0.16	0.14	0.1	0.61	0.60	0.56	0.56	0.05	0.03	0.03	0.04	0.10	0.08	0.08	0.11
Organic chemicals	0.17	0.19	0.18	0.18	0.53	0.53	0.51	0.51	0.11	0.12	0.13	0.13	0.26	0.28	0.28	0.28
Inorganic chemicals	0.08	0.09	0.09	0.09	0.19	0.19	0.18	0.18	0.03	0.02	0.02	0.02	0.20	0.13	0.09	0.05
Fertilizers	0.12	0.13	0.13	0.13	0.30	0.30	0.28	0.28	0.04	0.02	0.02	0.03	0.12	0.08	0.06	0.07
Chemical products industry	0.14	0.16	0.16	0.15	0.38	0.38	0.36	0.36	0.06	0.03	0.03	0.05	0.33	0.22	0.16	0.13
Other chemical industry	0.14	0.16	0.16	0.15	0.38	0.38	0.36	0.36	0.06	0.03	0.03	0.05	0.34	0.22	0.17	0.13
Iron and steel	0.03	0.04	0.04	0.04	0.08	0.08	0.07	0.07	0.01	0.02	0.04	0.05	0.08	0.08	0.09	0.09
Non-ferrous metals	0.01	0.02	0.03	0.03	0.04	0.06	0.05	0.05	0.05	0.03	0.05	0.05	0.06	0.07	0.11	0.11
Metal products	0.09	0.10	0.10	0.10	0.24	0.23	0.21	0.21	0.07	0.03	0.04	0.06	0.23	0.19	0.13	0.15
Building materials	0.12	0.14	0.13	0.12	0.36	0.37	0.35	0.34	0.11	0.06	0.09	0.10	0.16	0.13	0.32	0.36
Other industry	0.09	0.11	0.11	0.10	0.27	0.27	0.25	0.25	0.07	0.05	0.05	0.06	0.09	0.07	0.13	0.14
Construction	0.07	0.05	0.04	0.03	0.11	0.08	0.05	0.05	0.03	0.02	0.02	0.03	0.12	0.05	0.05	0.05
Services	0.15	0.16	0.15	0.14	0.50	0.46	0.42	0.42	0.05	0.03	0.03	0.04	0.14	0.11	0.08	0.09
Government	0.15	0.16	0.15	0.14	0.49	0.43	0.39	0.38	0.07	0.06	0.07	0.08	0.12	0.09	0.17	0.18
Transport	0.03	0.03	0.03	0.02	0.07	0.07	0.05	0.04	0.00	0.00	0.00	0.00	0.00	0.00	0.00	0.00
Households	0.13	0.14	0.14	0.13	0.34	0.31	0.29	0.29	0.09	0.09	0.11	0.12	0.23	0.15	0.27	0.29

Table 7.B.4b *Effects of a fast and slow speed of adjustment for retrofit on energy use under asymmetric S-curve*

	Electricity								Fuel							
	Fast adjustment				Slow adjustment				Fast adjustment				Slow adjustment			
	2005	2010	2015	2020	2005	2010	2015	2020	2005	2010	2015	2020	2005	2010	2015	2020
% change compared to baseline																
Horticulture	0.47	0.43	0.40	0.43	0.48	0.58	0.57	0.53	0.22	0.20	0.40	0.46	0.29	0.23	0.36	0.50
Other Agriculture	0.56	0.59	0.64	0.70	0.50	0.63	0.70	0.76	0.36	0.18	0.09	0.09	0.10	0.05	0.04	0.08
Food & beverage industry	1.22	1.34	1.45	1.58	0.86	1.38	1.77	2.06	0.41	0.33	0.30	0.21	0.31	0.22	0.15	0.16
Textile industry	0.24	0.25	0.26	0.30	0.16	0.26	0.33	0.4	0.55	0.44	0.36	0.29	0.11	0.09	0.07	0.06
Paper industry	1.63	1.85	2.10	2.50	1.29	1.96	2.43	2.93	0.16	0.12	0.16	0.24	0.19	0.09	0.12	0.20
Organic chemicals	1.20	1.35	1.49	1.67	1.00	1.41	1.72	2.01	0.36	0.42	0.45	0.48	0.28	0.35	0.41	0.47
Inorganic chemicals	0.27	0.29	0.29	0.31	0.19	0.25	0.29	0.33	0.23	0.15	0.11	0.06	0.04	0.02	0.02	0.03
Fertilizers	0.59	0.68	0.74	0.83	0.49	0.71	0.87	1.02	0.13	0.09	0.07	0.09	0.06	0.03	0.03	0.06
Chemical products industry	0.70	0.77	0.82	0.90	0.56	0.77	0.92	1.06	0.38	0.25	0.19	0.17	0.13	0.07	0.06	0.12
Other chemical industry	0.70	0.77	0.82	0.90	0.56	0.77	0.92	1.06	0.39	0.25	0.20	0.17	0.12	0.07	0.06	0.11
Iron and steel	0.09	0.10	0.09	0.09	0.04	0.06	0.07	0.08	0.09	0.11	0.13	0.14	0.03	0.05	0.10	0.13
Non-ferrous metals	0.05	0.08	0.08	0.08	0.03	0.06	0.08	0.09	0.08	0.08	0.15	0.16	0.09	0.07	0.10	0.13
Metal products	0.39	0.41	0.43	0.48	0.29	0.40	0.48	0.57	0.27	0.22	0.19	0.25	0.17	0.09	0.13	0.20
Building materials	0.70	0.81	0.88	1.01	0.53	0.79	0.99	1.17	0.36	0.26	0.74	0.97	0.62	0.33	0.56	0.89
Other industry	0.47	0.52	0.56	0.63	0.33	0.50	0.63	0.76	0.13	0.11	0.22	0.26	0.18	0.13	0.18	0.24
Construction	0.19	0.14	0.10	0.10	0.22	0.19	0.15	0.13	0.13	0.07	0.06	0.06	0.05	0.04	0.05	0.05
Services	1.16	1.21	1.27	1.41	0.99	1.33	1.51	1.66	0.17	0.14	0.14	0.19	0.16	0.09	0.12	0.19
Government	1.11	1.11	1.14	1.25	0.98	1.26	1.38	1.49	0.20	0.14	0.30	0.35	0.23	0.20	0.26	0.35
Transport	0.18	0.2	0.17	0.18	0.15	0.20	0.18	0.18	0.00	0.00	0.00	0.00	0.00	0.00	0.00	0.00
Households	1.08	1.16	1.27	1.44	1.00	1.36	1.54	1.68	0.51	0.32	0.62	0.74	0.56	0.53	0.68	0.89

NOTES

1. In this respect, our analysis differs from Jacobsen (2000), who examines technology diffusion in applied economic vintage models for Denmark, focusing on electricity and heat production, and residential demand for heating and electric appliances.
2. Fuel represents different types of oil products, gas, steam and coal.
3. Of course, this log-linear equation could also have been obtained by starting out with Cobb-Douglas production functions. However the makers of NEMO found the Cobb-Douglas structure too restrictive. Moreover a Cobb-Douglas formulation would have yielded equations for E/Z which would not be log-linear in P_E/P_Z, if P_H is expressed as a weighted average of P_E and P_I (see Koopmans and te Velde 2001).
4. Recall also that the underlying database only provides information about additional capital investments necessary to save energy. Relative energy intensity is defined as $F_{\tau,t} = F_{\tau,t} / F_{t0}$ with $F_{t0} = 0$.
5. For presentational purposes, Figure 7.3 assumes $a_{max} = 10$.
6. So aggregate capital is defined as $K_t = \left[\sum_{\tau=t-a_{max}}^{t} \left(K_{\tau,t}\right)^{\frac{\varepsilon-1}{\varepsilon}} \right]^{\frac{\varepsilon}{\varepsilon-1}}$
7. Assuming a lifetime interval of $[a_{min}, a_{max}] = [10,18]$.
8. Standard complementarity is defined by an elasticity of substitution $\varepsilon = 3$
9. Lijesen et al. (2001) use NEMO to analyse the effect of changes in the Dutch energy tax system (*Regulerende Energiebelasting*) for different scenarios. The marginal energy price increase in their study ranges from 0.1 per cent to 9 per cent, depending on the sector and the scenario. Hence, our uniform increase in energy price by 5 per cent can be seen as an average, in order to analyse the overall impact of an energy price increase in the context of different assumptions with respect to technology diffusion processes.
10. High and low complementarity are defined by an elasticity of substitution of $\varepsilon = 1.5$ and $\varepsilon = 15$ respectively.
11. Assuming a lifetime interval of $[a_{min}, a_{max}] = [10,18]$
12. Details are available upon request.
13. We refer to Tables 7.B.4a and 7.B.4b for the calculated differences in electricity and fuel consumption per sector, for a fast and slow diffusion speed for both S-curves.

8. Conclusions

8.1 MAIN CONCLUSIONS

In this book we have analysed several aspects of the interplay between economic growth, energy use and technological change. The book started with the presumption that technological change plays a crucial role in realizing energy efficiency improvements, and hence in ameliorating the conflict between economic growth and environmental quality. The principal focus of the research reported in this book has been the adoption and diffusion of energy-saving technologies, with much emphasis on the interaction between microeconomic investment decisions and macroeconomic patterns of energy-saving and productivity improvements. We analysed this issue from both a theoretical and an empirical perspective, and concluded with an assessment of the relevance of several of the obtained insights for economic policy analysis of sectoral energy-demand.

The research in this book has been inspired by recent developments in economic thinking about economic growth and technological change. Dissatisfaction with the old neoclassical theories of economic growth have led to new theories of economic growth and technological change, most notably the so-called endogenous growth theory and the evolutionary economic theory. Their main contribution consists of providing a more adequate view on technological change, moving beyond the unsatisfactory feature of old neoclassical growth models that assumes technological change as an exogenous process. Both new classes of models have endogenized technological change within, respectively, a neoclassical framework and according to an evolutionary view of economic growth and technological change. In Part I we reviewed these developments in search of the essential differences and similarities between the two approaches. We concluded that although evolutionary models often include a more sophisticated representation of technological and behavioural diversity, uncertainty, path dependency and irreversibility, they do not always clearly expose the mechanisms behind aggregate dynamics due to their relatively high degree of complexity. Moreover we argued that although evolutionary modelling has evolved since the seminal work by Nelson and Winter (1982), evolutionary growth theory has demonstrated relatively little progress in terms of formal theorizing. Finally, we argued that neoclassical and evolutionary models of technological change tend to converge in what is called a Schumpeterian view on economic growth and technological change, emphasizing the occurrence of a succession of innovations as the result of intended R&D activities, implying a business stealing effect ('creative destruction'). In our opinion, this argues for a combination of the insights of both approaches to improve our understanding of the complex processes of technological change in relation to other macro- and microeconomic processes.

In Part II we developed two theoretical models on adoption and diffusion of energy-saving technologies, providing a closer look at the energy efficiency paradox. Building upon the above-mentioned developments in recent economic theorizing about technological change, these models put forward the role of knowledge accumulation (learning), knowledge spillovers, uncertainty and path dependency in explaining why diffusion of energy-saving technologies is often so costly, difficult, and hence slow. In the vintage model developed in Chapter 3, we offered a new explanation for the energy efficiency paradox by arguing that the existence of technological complementarities provides an incentive for firms to continue to invest in old technologies, even if new energy-saving technologies are available that seem to be 'better' when considered in isolation. Moreover we showed that this investment barrier becomes larger if these 'returns to diversity' are accompanied by a learning-by-using effect, since this implies that a firm faces a loss of expertise on a particular vintage when it switches to a newer vintage. The model developed in Chapter 4 analysed the impact of microeconomic investment behaviour on aggregate energy savings, under technological uncertainty and in the presence of learning effects. We showed that neglecting the effects of uncertainty on investment behaviour results in an overestimation of short-run energy savings and an underestimation of long-run savings. Moreover we argued that investment subsidies have an ambiguous effect on aggregate energy saving. Although they stimulate energy savings in the short run, they may contribute to a lock-in into relatively inferior technologies, and hence can be counter-effective in the long run.

In Part III we moved from theory to empirics by analysing trends in energy and labour productivity performance across countries and across sectors. We traced back macroeconomic productivity developments to developments at the level of individual sectors, in order to correct as much as possible for the impact of structural effects on productivity trends. We started in Chapter 5 by documenting several stylized facts on energy and labour productivity trends at different levels of aggregation. Moreover, we performed a cross-country decomposition analysis of productivity growth rates, which revealed that structural changes contribute considerably to aggregate energy productivity growth in some countries, while they partly offset energy efficiency improvements in other countries. In contrast, we found structural changes to play only a minor role in explaining aggregate labour productivity developments. Furthermore we found labour productivity growth to be higher on average than energy productivity growth, and that this bias towards labour productivity growth has increased in the transport, agriculture and aggregate manufacturing sectors, while it has decreased in services and most manufacturing sectors.

The recent introduction of endogenous growth models not only gave rise to many theoretical studies on technological change, but also invoked new empirical studies on a central issue in the growth literature, namely that of convergence of income and productivity among countries. Inspired by this renewed interest in empirical convergence analysis, Chapter 6 provided a systematic comparison of energy and labour productivity convergence at a detailed sector level – whereas most studies focus on convergence of income, labour or total factor productivity at an aggregate level. We found the development of cross-country variation in energy and labour productivity performance to depend on the level of aggregation, with different patterns of convergence and divergence across sectors. Using a panel-data approach, we found evidence to suggest that in most sectors energy productivity grows relatively fast in countries with relatively low initial productivity levels, while in several sectors this appeared to be true also in terms of labour productivity. Hence our research provided support for the hypothesis that lagging countries tend to catch up with technological leaders. Nevertheless, at the same time it turned out that in virtually all sectors there remain substantial cross-country productivity differences, in particular in terms of energy productivity. Further analyses revealed convergence to be conditional rather than unconditional, suggesting productivity levels to converge not so much to a uniform level but rather to country-specific steady states. Finally, we searched in Chapter 6 for the fundamentals determining these country-specific steady states. We found energy prices to stimulate energy productivity growth in the energy intensive sectors, whereas we did not find much evidence of a positive relationship between wages and labour productivity growth. Moreover we found economies of scale to have contributed to energy and labour productivity growth in several sectors, while investment share, openness and specialization played only a very limited role in explaining (cross-country differences in) energy and labour productivity growth.

In Part IV we integrated the theoretical insights developed in Part II into NEMO, an existing applied economic model of economy–energy interaction used for policy analysis in the Netherlands. We found that the inclusion of a rather detailed description of technology diffusion processes in a macroeconomic model leads to higher estimates of energy demand as compared to most macroeconomic models including endogenous innovation and learning-by-doing processes. The reason is that the latter type of model usually ignores the discontinuities and temporary loss of expertise inherent to microeconomic processes of technological change. These findings support the relevance of including technology diffusion in applied economic policy models.

8.2 POLICY ISSUES

Most governments in the developed world strive explicitly for sustainable economic growth, aiming to decouple economic growth and environmental pressure. Under the United Nations Framework Convention on Climate Change (UNFCCC), governments of most industrialized countries have committed themselves to substantially reduce their energy-related greenhouse gas emissions within a specified time horizon. Increasing energy efficiency through technological progress is considered to be an important option to reach this goal. Hence many governments face the questions of how and when to stimulate the adoption of energy-saving technologies. Although this book has not been written to provide a blueprint or clear recipe for policymakers, a few suggestions can be made concerning policies to stimulate energy efficiency.

The rationale for government intervention to increase energy efficiency lies in the existence of two market failures. First, ever increasing energy demand brings about environmental externalities, of which the enhanced greenhouse effect features prominently. Second, because of appropriability and spillover effects inherent in processes of technological change, a laissez-faire world tends to generate insufficient energy-saving technological progress in terms of what is socially optimal, since private firms do not internalize positive social externalities (see also Chapter 2). Consequently government intervention should address these market failures in, respectively, two ways. First, through environmental policies, including measures to make the polluters pay and to ensure that prices reflect social costs (for example via energy taxes). Second, through technology policies aiming to address 'technology market failures', including subsidies to stimulate R&D and the adoption of technologies. If environmental policies address the environmental externalities adequately, general technology policies will suffice to stimulate energy-saving technological change. In other words there is no need for specific energy-saving technology policies since the internalisation of environmental externalities will trigger energy-augmenting technological change via relative factor prices. Of course reality does not meet this presumption, and hence governments do have a rationale to engage in policies to stimulate the adoption of energy-saving policies. In the meantime however they should not forget to strive for a first-best solution.

In a second-best world, governments should shape energy-saving technology policies with caution. In Chapter 4 we have shown that investment subsidies may be counter-effective since they can contribute to the lock-in into relatively inferior technologies. We have argued that this requires subsidy programmes to take into account rather detailed information on the nature of technological change, including the timing of innovations, learning effects, and the quality (energy-saving potential) of new technologies. Governments are unlikely to have all this information, implying that subsidy programmes have a high risk of being ineffective and of inducing free-rider effects.

Throughout this book we have illustrated the role of knowledge accumulation in technology diffusion processes. Since learning effects, triggered by the adoption and use of new technologies, lead to improvement in the quality of technologies, spillovers of knowledge and/or decreasing costs, government intervention should aim to contribute to these learning processes. Moreover we have argued that the diffusion of technologies is at least as important, costly and difficult as the development of new technologies. Not only because it is obvious that energy efficiency improvements will not be realized with innovation alone, but also because it is known that the widespread adoption of existing energy-saving technologies could enable a significant reduction in energy use, especially in the short and medium run.

Hence the allocation of government expenditures on stimulating energy-saving technological change should be the right mix of support to R&D, market introduction ('demonstration') and large-scale adoption ('commercialization'). Of course it is extremely difficult to say in general terms what 'right' is in this context. Nevertheless the considerable energy-saving potential of existing new technologies, and the presence of learning effects, both underline the importance of market introduction and diffusion processes in realizing energy efficiency improvements. Hence these phases of technological change should also receive sufficient attention in terms of allocation of government expenditures. At the same time however learning effects provide an argument in favour of limited, well-focused and temporary investment subsidies aiming to trigger learning effects. Once the scale of adoption is such that new technologies are competitive and learning effects become small, there is no need to continue subsidizing their adoption (for a further elaboration on this argument see OECD/IEA 2000).

8.3 FUTURE RESEARCH

Of course, more research is needed. In the remaining part of this chapter, we will briefly mention a few issues that might deserve particular attention following the research presented in this book.

In Chapter 3 we developed a simple vintage model in which technological change was characterized by returns to using a mix of different vintages and a learning-by-using effect. In particular due to its vintage character and its rather detailed description of microeconomic investment behaviour in a macroeconomic context, the model is well suited for analysing the energy efficiency paradox. The model could be extended in a number of interesting directions. First, one could allow for the endogenous determination of the rate of learning-by-using, the rate of improvement of new vintages and the increase in energy efficiency. We refer here to Aghion and Howitt (1996) for this kind of analysis, drawing a distinction between research (developing new vintages) and development (improving existing vintages). Second, one could allow for the incomplete depreciation of capital in order to assess the importance of complementarity in understanding the development of the stock of capital of different vintages and the investment behaviour of firms. Third, the production structure could be modelled in a more realistic way by allowing for substitution between energy and capital.

Although energy is an essential factor for economic growth it is not the only input in production processes. Hence, energy demand depends not only on the level of output and the degree of technological progress, but also on the substitution possibilities among the different inputs, their relative prices and the nature of technological change. In this book we have provided a first glance at the role of these factors in determining energy demand. In Chapter 5 we have analysed trends in energy and labour productivity, including an examination of the relation between energy and labour productivity growth rates. In Chapter 6, we have searched for factors determining energy and labour productivity growth, including factor prices. However to understand the relative importance of the impact of technological change, input substitution possibilities and relative factor prices in determining energy demand, it is necessary to go one step further in the analysis by empirically estimating production functions. Such an analysis would provide valuable information on energy–capital–labour substitution, which is an important issue concerning the interplay between economic growth, energy use and technological change, because the economic effects of energy or climate change policies aiming to increase energy efficiency and/or reduce greenhouse gas emissions depends critically on the degree to which capital and labour can be substituted for energy. This line of research should address the considerable uncertainty and disagreement in the literature, (1) as to which type of production function is most suitable to describe the production behaviour of the economy (see for example Diewert and Wales 1987, Gagné and Quellette 2002) and (2) with respect to the numerical values of the substitution elasticities (see for example Kemfert 1998).

In Chapter 6 we tried to identify the variables that determine energy and labour productivity growth in different countries. We found energy prices, wages and economies of scale to stimulate respectively energy and labour productivity growth in various sectors while investment share, openness and specialization seem to play only a very limited role in explaining (cross-country differences in) energy and labour productivity growth. At the same time, however, we found unspecified country-effects to be an important factor in explaining energy and labour productivity growth among countries. As argued in Chapter 6, this suggests a need for additional variables, in order to further explain sectoral trends in energy and labour productivity growth across countries. Following recent developments in economic theorizing of economic growth and technological change, R&D and human capital are promising candidates in this respect.

Since productivity growth is primarily driven by technological change, cross-country productivity differences suggest the existence of different technology levels among countries. The productivity convergence analysis presented in Chapter 6 provided insight into the development of energy and labour productivity differences over time, and hence in possible patterns of international technology flows. We concluded that convergence is conditional rather than unconditional and that in spite of evidence of convergence in most sectors, there remain substantial cross-country productivity differentials, particularly in terms of energy productivity. This suggests that technological spillovers are at least to some degree local, a hypothesis that has been confirmed by Keller (2002) through empirical research of R&D spillovers within 14 OECD countries. It would be interesting to further assess the issue of conditional convergence by examining the role of international technology diffusion in determining cross-country productivity differentials, including non-OECD countries. A starting point for such an analysis can be found in Gomulka (1971) who found a hump-shaped pattern between economic growth and the technology gap (distance to technology leader). This suggests that regions or countries lagging too far behind are not able to catch up with the leading country, a suggestion that is much in line with Abramovitz's (1986) notion that 'social capabilities' are required in order for successful catching up. Chatterji (Chatterji 1992, Chatterji et al. 1993) took up this idea and confirmed the hypothesis that different sets of countries are converging to different steady-state levels, with backward countries staying behind. A better view of international patterns of energy productivity convergence clubs is needed to better understand the causes behind technological gaps, and hence the role technology diffusion can play in increasing energy efficiency in different world regions. This is important, since decoupling of economic growth and environmental problems has become a global issue.

References

Abramovitz, M. (1986), 'Catching Up, Forging Ahead, and Falling Behind', *Journal of Economic History*, **47**, 385–406.

Acemoglu, D. (2002), 'Directed Technical Change', *Review of Economic Studies*, **69**, 781–809.

Adler, P.S. and K.B. Clark (1991), 'Behind the Learning Curve: A Sketch of the Learning Process', *Management Science*, **37**, 267–81.

Aghion, P. and P. Howitt (1992), 'A Model of Growth through Creative Destruction', *Econometrica*, **60**, 323–51.

Aghion, P. and P. Howitt (1996), 'Research and Development in the Growth Process', *Journal of Economic Growth*, **1**, 49–73.

Aghion, P. and P. Howitt (1998), *Endogenous Growth Theory*, Cambridge, MA: MIT Press.

Aghion, P., M. Dewatripont and P. Rey (1997), 'Corporate Governance, Competition Policy and Industrial Policy', *European Economic Review*, **41**, 797–805.

Aghion, P., M. Dewatripont and P. Rey (1999), 'Competition, Financial Discipline and Growth', *Review of Economic Studies*, **66**, 825–52.

Alvarez, F. and H. Amman (1999), 'Learning-by-Doing under Uncertainty', *Computational Economics*, **14**, 255–62.

Ang, B.W. (1995a), 'Decomposition Methodology in Industrial Energy Demand Analysis', *Energy*, **20**, 1081–95.

Ang, B.W. (1995b), 'Multilevel Decomposition of Industrial Energy Consumption', *Energy Economics*, **17**, 39–51.

Ang, B.W. (1999), 'Decomposition Methodology in Energy Demand and Environmental Analysis', in J.C.J.M. van den Bergh (ed.), *Handbook of Environmental and Resource Economics*, Cheltenham: Edward Elgar, pp. 1146–63.

Ang, B.W. and F.Q. Zhang (2000), 'A Survey of Index Decomposition Analysis in Energy and Environmental Studies', *Energy*, **25**, 1149–76.

Antonelli, C. (1993), 'Externalities and Complementarities in Telecommunications Dynamics', *Journal of Industrial Organisation*, **11**, 437–48.

Antonelli, C., P. Petit and G. Tahar (1990), 'The Diffusion of Interdependent Innovations in the Textile Industry', *Structural Change and Economic Dynamics*, **1**, 207–21.

Argote, L. and D. Epple (1990), 'Learning Curves in Manufacturing', *Science*, **247**, 920–24.

Ark, B. van (1993), *International Comparisons of Output and Productivity*, Monograph Series no. 1, Groningen: Growth and Development Centre.

Ark, B. van (1996), 'Sectoral Growth Accounting and Structural Change in Post-War Europe', in B. van Ark and N. Crafts (eds), *Quantitative Aspects of Post-War*

European Economic Growth, Cambridge: Cambridge University Press, pp. 84–164.

Ark, B. van (1997), *Economic Growth in the Long Run: A History of Empirical Evidence*, International Library of Critical Writings in Economics 76 (3 Vol), Cheltenham: Edward Elgar.

Ark, B. van and N. Crafts (eds) (1996), *Quantitative Aspects of Post-War European Growth*, Cambridge: Cambridge University Press.

Ark, B. van and D. Pilat (1993), 'Productivity Level in Germany, Japan, and the United States: Differences and Causes', *Brookings Papers on Economic Activity: Microeconomics*, **2**, 1–46

Arrow, K.J. (1962), 'The Economic Implications of Learning-by-Doing', *Review of Economic Studies*, **29**, 155–73.

Arthur, W.B. (1989), 'Competing Technologies, Increasing Returns, and Lock-in By Historical Events', *Economic Journal*, **99**, 116–31.

Balcer, Y. and S.A. Lippman (1984), 'Technological Expectations and Adoption of Improved Technology', *Journal of Economic Theory*, **34**, 292–318.

Barro, R.J. (1991), 'Economic Growth in a Cross-Section of Countries', *Quarterly Journal of Economics*, **106**, 407–43.

Barro, R.J. (1997), *Determinants of Economic Growth: A Cross-Country Empirical Study*, Cambridge, MA: MIT Press.

Barro, R.J. and X. Sala-i-Martin (1992), 'Convergence', *Journal of Political Economy*, **100**, 223–51.

Barro, R.J. and X. Sala-i-Martin (1995), *Economic Growth*, New York: McGraw Hill.

Barro, R.J., N.G. Mankiw and X. Sala-i-Martin (1995), 'Capital Mobility in Neoclassical Models of Growth', *American Economic Review*, **85**, 103–15.

Baumol, W.J. (1986), 'Productivity Growth, Convergence and Welfare: What the Long-Run Data Show', *American Economic Review*, **76**, 1072–85.

Baumol, W.J., R.R. Nelson and E.N. Wolff (1994), *Convergence and Productivity; Cross-National Studies and Historical Evidence*, Oxford: Oxford University Press.

Beer, J. de (1998), *Potential for Industrial Energy-Efficiency*, PhD Thesis, Utrecht: Utrecht University.

Benkard, C.L. (2000), 'Learning and Forgetting: The Dynamics of Aircraft Production', *American Economic Review*, **90**, 1034–54.

Berg, S.V. and P. Friedman (1977), 'Joint Ventures, Competition and Technological Complementarities: Evidence from Chemicals', *Southern Economic Journal*, **43**, 1330–37.

Bergh, J.C.J.M. van den and J.M. Gowdy (2000), 'Evolutionary Theories in Environmental and Resource Economics: Approaches and Applications', *Environmental and Resource Economics*, **17**, 37–57.

Bernard, A.B. and S.N. Durlauf (1996), 'Interpreting Tests of the Convergence Hypothesis', *Journal of Econometrics*, **71**, 161–73.

Bernard, A.B. and C.I. Jones (1996a), 'Comparing Apples to Oranges: Productivity Convergence and Measurement across Industries and Countries', *American Economic Review*, **86**, 1216–38.

Bernard, A.B. and C.I. Jones (1996b), 'Productivity Across Industries and Countries: Time Series Theory and Evidence', *Review of Economics and Statistics*, **78**, 135–46.

Bernard, A.B. and C.I. Jones (2001), 'Comparing Apples to Oranges: Productivity Convergence and Measurement Across Industries and Countries: Reply', *American Economic Review*, **91**, 1168–69.

Berndt, E.R. (1978), 'Aggregate Energy, Efficiency, and Productivity Measurement', *Annual Review of Energy*, **3**, 225-73.

Berndt, E.R. and D.O. Wood (1975), 'Technology, Prices, and the Derived Demand for Energy', *Review of Economics and Statistics*, **57**, 259–68.

Berndt, E.R. and D.O. Wood (1979), 'Engineering and Econometric Interpretations of Energy Capital Complementarity', *American Economic Review*, **69**, 342–54.

Binswanger, H.P. (1974a), 'The Measurement of Technical Change Biases with Many Factors of Production', *American Economic Review*, **64**, 964–76.

Binswanger, H.P. (1974b), 'A Microeconomic Approach to Induced Innovation', *Economic Journal*, **84**, 940–58.

Blanchard, O.J. and S. Fisher (1989), *Lectures on Macroeconomics*, Cambridge, MA: MIT Press.

Böhringer, C. (1998), 'The Synthesis of Bottom-Up and Top-Down in Energy Policy Modeling', *Energy Economics*, **20**, 233–48.

Broadberry, S. N. (1996), 'Convergence: What the Historical Record Shows', in: B. van Ark and N. Crafts (eds), *Quantitative Aspects of Post-War European Growth*, Cambridge: Cambridge University Press, pp. 327–46.

Buonanno, P., C. Carraro and M. Galeotti (2003), 'Endogenous Induced Technical Change and the Costs of Kyoto', *Resource and Energy Economics*, **25**, 11–34.

Butter, F.A.G. den, R.B. Dellink and M.W. Hofkes (1995), 'Energy Levies and Endogenous Technology in an Empirical Simulation Model for the Netherlands', in: A.L. Bovenberg and S. Cnossen (eds), *Public Economics and the Environment in an Imperfect World*, Boston, MA: Kluwer Academic Publishers, pp. 315–35.

Canton, E.J.F., H.L.F. de Groot and R. Nahuis (2002), 'Vested Interests, Population Ageing and Technology Adoption', *European Journal of Political Economy*, **18**, 631–52.

Carraro, C., R. Gerlagh and B. van der Zwaan (2003), 'Endogenous Technical Change in Environmental Macroeconomics', *Resource and Energy Economics*, **25**, 1–10.

Chari, V.V. and H. Hopenhayn (1991), 'Vintage Human Capital, Growth, and the Diffusion of New Technology', *Journal of Political Economy*, **99**, 1142–65.

Chatterji, M. (1992), 'Convergence Clubs and Endogenous Growth', *Oxford Review of Economic Policy,* **8**, 57–69.

Chatterji, M., B. Gilmore, K. Strunk and J. Vanasin (1993), 'Political Economy, Growth and Convergence in Less-Developed Countries', *World Development,* **21**, 2029–38.

Chiaromonte, F. and G. Dosi (1993), 'Heterogeneity, Competition, and Macroeconomic Dynamics', *Structural Changes and Economic Dynamics*, **4**, 39–63.

Choi, J.P. (1994), 'Irreversible Choice of Uncertain Technologies with Network Externalities', *RAND Journal of Economics*, **25**, 382–401.

Cimoli, M. and G. Dosi (1995), 'Technological Paradigms, Patterns of Learning and Development: An Introductory Roadmap', *Journal of Evolutionary Economics*, **5**, 243–68.

Clarke, L.E. and J.P. Weyant (2002), 'Modelling Induced Technical Change: An Overview', in A. Grübler, N. Naki•enovi• and W.D. Nordhaus, *Modelling Induced Innovation in Climate Change Policy*, Washington, DC: Resources for the Future.

Cohen, W. and D. Levinthal (1989), 'Learning and Innovation: The Two Faces of R&D', *Economic Journal*, **99**, 569–596.

Colombo, M.G. and R. Mosconi (1995), 'Complementarity and Cumulative Learning Effects in the Early Diffusion of Multiple Technologies', *Journal of Industrial Economics*, **43**, 13–48.

Conlisk, J. (1989), 'An Aggregate Model of Technical Change', *Quarterly Journal of Economics*, **104**, 787–821.

CPB (2000), *Naar een Efficiënter Milieubeleid*, Den Haag: Sdu Uitgevers.

Dalgaard, C.-J. and J. Vastrup (2001), 'On the Measurement of σ–Convergence', *Economics Letters*, **70**, 283–87.

David, P.A. (1975), *Technical Choice, Innovation and Economic Growth*, Cambridge: Cambridge University Press.

David, P.A. (1985), 'Clio and the Economics of QWERTY', *American Economic Review*, **75**, 332–37.

Davies, S. (1979), *The Diffusion of Process Innovations*, Cambridge: Cambridge University Press.

Decanio, S.J. and J.A. Laitner (1997), 'Modeling Technological Change in Energy Demand Forecasting. A Generalized Approach', *Technological Forecasting and Social Change*, **55**, 249–63.

DeLong, J.B. (1988), 'Productivity Growth, Convergence and Welfare: Comment', *American Economic Review*, **78**, 1138–54.

Denison, E.F. (1967), *Why Growth Rates Differ*, Washington, DC: Brookings.

Diewert, W.E. and T.J. Wales (1987), 'Flexible Functional Forms and Global Curvature Conditions', *Econometrica*, **55**, 43–68.

Dixit, A.K. and R.S. Pindyck (1994), *Investment under Uncertainty*, Princeton, NJ: Princeton University Press.

Dixit, A.K. and J.E. Stiglitz (1977), 'Monopolistic Competition and Optimum Product Diversity', *American Economic Review*, **67**, 297–308.

Dixon, R. (1980), 'Hybrid Corn Revisited', *Econometrica*, **48**, 1451–61.

Dollar, D. and E.N. Wolff (1988), 'Convergence of Industry Labour Productivity among Advanced Economies, 1963–1982', *Review of Economics and Statistics*, **70**, 549–58.

Dollar, D. and E.N. Wolff (1993), *Competitiveness, Convergence, and International Specialization*, Cambridge, MA: MIT Press.

Dosi, G. (1982), 'Technological Paradigms and Technological Trajectories', *Research Policy*, **11**, 147–62.

Dosi, G. (1988), 'Sources, Procedures and Microeconomic Effects of Innovation', *Journal of Economic Literature*, **26**, 1120–71.

Dosi, G. (1997), 'Opportunities, Incentives and the Collective Patterns of Technological Change', *Economic Journal*, **107**, 1530–47.

Dosi, G. and R. Nelson (1994), 'An Introduction to Evolutionary Theories in Economics', *Journal of Evolutionary Economics*, **4**, 153–72 .

Dosi, C. and M. Moretto (1997), 'Pollution Accumulation and Firm Incentives to Accelerate Technological Change under Uncertain Private Benefits', *Environmental and Resource Economics*, **10**, 285–300.

Dosi, G., C. Freeman, R.R. Nelson, G. Silverberg and L. Soete (1988), *Technical Change and Economic Theory*, London: Pinter Publishers.

Dosi, G., S. Fabiani and M. Meacci (1994), 'The Dynamics of International Differentiation: A Multi-Country Evolutionary Model', *Industrial and Corporate Change*, **3**, 225–41.

Durlauf, S.N. and P.A. Johnson (1992), *Local versus Global Convergence Across National Economies*, NBER Working Paper no. 3996, Cambridge, MA: NBER.

Durlauf, S.N. and D. Quah (1999), 'New Empirics of Economic Growth', in J.B. Taylor and M. Woodford (eds), *Handbook of Macroeconomics*, Volume 1, Amsterdam: North Holland Elsevier Science, pp. 235–308.

Economic Journal (1996), *Controversy on the Convergence and Divergence of Growth Rates*, Special Issue, **106**, 1016–69.

Edenhofer, O. and C.C. Jaeger (1998), 'Power Shifts: The Dynamics of Energy Efficiency', *Energy Economics*, **20**, 513–37.

Eichhammer, W. and W. Mannsbart (1997), 'Industrial Energy Efficiency: Indicators for a European Cross-Country Comparison of Energy Efficiency in the Manufacturing Industry, *Energy Policy*, **25**, 759–72.

Ethier, W.J. (1982), 'National and International Returns to Scale in the Modern Theory of International Trade', *American Economic Review*, **72**, 389–405.

Fagerberg, J. (1994), 'Technology and International Differences in Growth Rates', *Journal of Economic Literature*, **32**, 1147–75.

Fagerberg, J. (2000), 'Technological Progress, Structural Change and Productivity Growth', *Structural Change and Economic Dynamics*, **11**, 393–411.

Farla, J.C.M. and K. Blok (2000), 'The Use of Physical Indicators for the Monitoring of Energy Intensity Developments in the Netherlands, 1980–1995', *Energy*, **25**, 609–38.

Farzin, Y.H., K.J.M. Huisman and P.M. Kort (1998), 'Optimal Timing of Technology Adoption', *Journal of Economic Dynamics and Control*, **22**, 779–99.

Fisher, R.A. (1930), *The Genetic Theory of Natural Selection*, Oxford: Clarendon Press.

Frankel, M. (1962), 'The Production Function in Allocation and Growth: A Synthesis', *American Economic Review*, **52**, 995–1022.

Freeman, C. (1994), 'The Economics of Technological Change', *Cambridge Journal of Economics*, **18**, 463–514.

Freeman, S.L., M.J. Niefer and J.M. Roop (1997), 'Measuring Industrial Energy Intensity: Practical Issues and Problems', *Energy Policy*, **25**, 703–14.

Fuss, M.A. (1977), 'The Structure of Technology over Time: A Model for Testing the 'Putty–Clay' Hypothesis', *Econometrica*, **45**, 1797–1821.

Gagné, R. and P. Quellette (2002), 'The Effect of Technological Change and Technical Inefficiencies on the Performance of Functional Forms', *Journal of Productivity Analysis*, **17**, 233–47.

Gerlagh, R. and B. van der Zwaan (2003), 'Gross World Product and Consumption in a Global Warming Model with Endogenous Technological Change', *Resource and Energy Economics*, **25**, 35–58.

Geroski, P.A. (2000), 'Models of Technology Diffusion', *Research Policy*, **29**, 603–25.

Gerschenkron, A. (1952), 'Economic Backwardness in Historical Perspective', in B. Hoselitz (ed.), *The Progress of Underdeveloped Areas*, Chicago, IL: University of Chicago Press, pp. 5–30.

Gomulka, S. (1971), *Inventive Activity, Diffusion, and the Stages of Economic Growth*, Monograph 24, Aarhus: Institute of Economics.

Gomulka, S. (1976), 'Do New Factories Embody Best Practice Technologies?', *Economic Journal*, **86**, 859–63

Goodwin, R.M. (1967), 'A Growth Cycle', in: C.H. Feinstein (ed.), *Socialism, Capitalism and Economic Growth*, Cambridge: Cambridge University Press, pp. 54–8.

Goulder, L.W. and K. Mathai (2000), 'Optimal CO_2, Abatement in the Presence of Induced Technological Change', *Journal of Environmental Economics and Management*, **39**, 1–38.

Goulder, L.W. and S.H. Schneider (1999), 'Induced Technological Change and the Attractiveness of CO_2 Abatement Policies', *Resource and Energy Economics*, **21**, 211–53.

Greening, L.A., W.B. Davis and L. Schipper (1997), 'Comparison of Six Decomposition Methods: Application to Aggregate Energy Intensity for Manufacturing in 10 OECD Countries', *Energy Economics*, **19**, 375–90.

Griliches, Z. (1957), 'Hybrid Corn: An Exploration of the Economics of Technical Change', *Econometrica*, **25**, 501–22.

Groenenberg, H. (2002), *Development and Convergence. A Bottom-Up Analysis for the Differentiation of Future Commitments under the Climate Convention*, PhD Thesis, Utrecht: Utrecht University.

Groot, H.L.F. de, M.W. Hofkes and P. Mulder (2003), 'A Vintage Model of Technology Diffusion: The Effects of Returns to Diversity and Learning-by-Using', in S. Brakman and B.J. Heijdra (eds), *The Monopolistic Competition Revolution in Retrospect*, Cambridge: Cambridge University Press, pp. 356–72.

Groot, H.L.F. de, P. Mulder and D.P. van Soest (2002), *Subsidising the Adoption of Energy-Saving Technologies. Analyzing the Impact of Uncertainty, Learning and Maturation*, OCfEB Research Memorandum no. 0201, Rotterdam: Erasmus University.

Grossman, G.M. and E. Helpman (1990), 'Comparative Advantage and Long-Run Growth', *American Economic Review*, **80**, 796–815.

Grossman, G.M. and E. Helpman (1991), *Innovation and Growth in the Global Economy*, Cambridge, MA: MIT Press.

Grubb, M. (1997), 'Technologies, Energy Systems and the Timing of CO2, Emissions Abatement. An Overview of Economic Issues', *Energy Policy*, **25**, 159–72.

Grübler, A. and S. Messner (1998), 'Technological Change and the Timing of Abatement Measures', *Energy Economics*, **20**, 495–512.

Grübler, A., N. Nakicenovic and D.G. Victor (1999), 'Dynamics of Energy Technologies and Global Change', *Energy Policy*, **27**, 247–80.

Hartog, H. den and H.S. Tjan (1980), 'A Clay–Clay Vintage Model Approach for Sectors of Industry in the Netherlands', *De Economist*, **128**, 129–88.

Helpman, E. (1992), 'Endogenous Macroeconomic Growth Theory', *European Economic Review*, **36**, 237–67.

Hicks, J.R. (1932), *The Theory of Wages* (2nd edn 1966), London: Macmillan.

Hirsch, W.Z. (1952), 'Progress Functions of Machine Tool Manufacturing', *Econometrica*, **20**, 81–82.

Hodgson, G.M. (1993), *Economics and Evolution*, Cambridge: Polity Press.

Hoppe, H.C. (2000), 'Second-Mover Advantages in the Strategic Adoption of New Technology under Uncertainty', *International Journal of Industrial Organization*, **18**, 315–38.

Howarth, R.B., L. Schipper, P.A. Duerr and S. Strøm (1991), 'Manufacturing Energy Use in 8 OECD Countries', *Energy Economics*, **13**, 135–42.

Hudson, E.A. and D.W. Jorgenson (1974), 'US Energy Policy and Economic Growth, 1975–2000', *Bell Journal of Economics and Management Science*, **5**, 461–514.

Interlaboratory Working Group (IWG) (1997), *Scenarios of Carbon Reductions: Potential Impacts of Energy Technologies by 2010 and Beyond*, Washington, DC:

Office of Energy Efficiency and Renewable Technologies, US Department of Energy.

Islam, N. (1995), 'Growth Empirics: A Panel Data Approach', *Quarterly Journal of Economics*, **4**, 1127–70.

Islam, N. (2003), 'What Have We Learnt from the Convergence Debate?', *Journal of Economic Surveys*, **17**, 309–62.

Jacobsen, H.K. (2000), 'Technology Diffusion in Energy-Economy Models: The Case of Danish Vintage Models', *Energy Journal*, **21**, 43–71.

Jaffe, A.B. and R.N. Stavins (1994), 'The Energy Paradox and the Diffusion of Conservation Technology', *Resource and Energy Economics*, **16**, 91–122.

Jaffe, A.B., R.G. Newell and R.N. Stavins (1999), *Energy-Efficient Technologies and Climate Change Policies: Issues and Evidence*, Climate Issue Brief 19, Washington, DC: Resources for the Future.

Jaffe, A.B., R.G. Newell and R.N. Stavins (2002), 'Environmental Policy and Technological Change', *Environmental and Resource Economics*, **22**, 41–69.

Jones, C.I. (1997), 'On the Evolution of World Income Distribution', *Journal of Economic Perspectives*, **11**, 19–36.

Jorgenson, D.W. (1984), 'The Role of Energy in Productivity Growth', *Energy Journal*, **5**, 11–26.

Jorgenson, D.W. (1986), 'The Great Transition: Energy and Economic Change', *Energy Journal*, **7**, 1–13.

Jorgenson, D.W. (1995), *Productivity, Vol 2: International Comparisons of Economic Growth*, Cambridge, MA: MIT Press.

Jorgenson, D.W. and Z. Griliches (1967), 'The Explanation of Productivity Change', *Review of Economic Studies*, **34**, 249–82.

Jovanovic, B. (1997), 'Learning and Growth', in D.M. Kreps and K.F. Wallis (eds), *Advances in Economics and Econometrics: Theory and Applications*, Vol. II, Cambridge: Cambridge University Press, pp. 318–39.

Jovanovic, B. and S. Lach (1989), 'Entry, Exit, and Diffusion with Learning by Doing', *American Economic Review*, **79**, 690–99.

Jovanovic, B. and Y. Nyarko (1996), 'Learning by Doing and the Choice of Technology', *Econometrica*, **64**, 1299–1310.

Jovanovic, B. and D. Stolyarov (2000), 'Optimal Adoption of Complementary Technologies', *American Economic Review*, **90**, 15–29.

Kaldor, N. (1961), 'Capital Accumulation and Economic Growth', in F.A. Lutz and D.C. Hague (eds), *The Theory of Capital*, London: Macmillan.

Kamien, M.I. and N.L. Schwartz (1972), 'Timing of Innovations under Rivalry', *Econometrica*, **40**, 43–60.

Kapur, S. (1995), 'Technological Diffusion with Social Learning', *Journal of Industrial Economics*, **42**, 173–95.

Keller, W. (2002), 'Geographic Localization of International Technology Diffusion', *American Economic Review*, **92**, 120–42.

Kemfert, C. (1998), 'Estimated Elasticities of a Nested CES Production Function Approach for Germany', *Energy Economics*, **20**, 249–64.

Kendrick, J.W. (1961), *Productivity Trends in the United States*, Princeton, NJ: Princeton University Press.

Kennedy, C. (1962), 'The Character of Improvements and Technical Progress', *Economic Journal*, **72**, 899–911.

Klundert, Th. v.d. and S. Smulders (1992), 'Reconstructing Growth Theory, a Survey', *De Economist*, **140**, 177–203.

Koopmans, C.C and D.W. te Velde (2001), 'Bridging the Energy-Efficiency Gap: Using Bottom–up Information in a Top-down Energy Demand Model', *Energy Economics*, **23**, 57–75.

Koopmans, C.C., D.W. te Velde and W. Groot (1999), *NEMO: Netherlands Energy Demand Model. A Top-down Model Based on Bottum-up Information*, Research Memorandum no 155, The Hague: CPB Netherlands Bureau for Economic Policy Analysis.

Krusell, P. and J.-V. Ríos-Rull (1996), 'Vested Interest in a Positive Theory of Stagnation and Growth', *Review of Economic Studies*, **63**, 301–29.

Kumar, S. and R.R. Russell (2002), 'Technological Change, Technological Catch-Up, and Capital Deepening: Relative Contributions to Growth and Convergence', *American Economic Review*, **92**, 527–48.

Kuper, G.H. and D.P. van Soest (2003), 'Path-Dependency and Input Substitution: Implications for Energy Policy Modelling', *Energy Economics*, **25**, 397–407.

Lekvall, P. and C. Wahlbin (1973), 'A Study of Some Assumptions Underlying Innovation Diffusion Functions', *Swedish Journal of Economics*, **75**, 362–77.

Lijesen, M., M. Mulder and M. Vromans (2001), *Fiscale Vergroening en Energie II. Economische Effecten van Verhoging en Verbreding van de Regulerende Energiebelasting*, CPB Document 006, The Hague: CPB Netherlands Bureau for Economic Policy Analysis.

Lucas, R.E. (1988), 'On the Mechanics of Economic Development', *Journal of Monetary Economics*, **22**, 3–42.

Maddison, A. (1952), 'Productivity in an Expanding Economy', *Economic Journal*, **62**, 584–94.

Maddison, A. (1991), *Dynamic Forces in Capitalist Development*, Oxford: Oxford University Press.

Maddison, A. (1999), *Monitoring the World Economy 1820–1992*, Paris: OECD Development Centre Studies.

Majd, S. and R.S. Pindyck (1989), 'The Learning Curve and Optimal Production Under Uncertainty', *RAND Journal of Economics*, **20**, 331–43.

Malcomson, J.M. (1975), 'Replacement and the Rental Value of Capital Equipment Subject to Obsolescence', *Journal of Economic Theory*, **10**, 24–41.

Malerba, F. (1992), 'Learning by Firm and Incremental Technical Change', *Economic Journal*, **102**, 845–859.

Mankiw, N.G., D. Romer and D.N. Weil (1992), 'A Contribution to the Empirics of Economic Growth', *Quarterly Journal of Economics*, **107**, 407–37.

Mansfield, E. (1968), *Industrial Research and Technological Innovation*, New York: W.W. Norton.

Massell, B.F. (1961), 'A Disaggregated View of Technical Change', *Journal of Political Economy*, **69**, 547–57.

McGrattan, E.R. (1998), 'A Defense of AK Growth Models', *Federal Reserve Bank of Minneapolis Quarterly Review*, **22**, 13–27.

Meijers, H. (1994), *On the Diffusion of Technologies in a Vintage Framework. Theoretical Considerations and Empirical Results*, PhD Thesis, Maastricht: Maastricht University.

Metcalfe, J.S. (1988), 'The Diffusion of Innovation: An Interpretative Survey', in G. Dosi et al. (eds), *Technical Change and Economic Theory*, London: Pinter Publishers, pp. 560–89.

Miketa, A. (2001), 'Analysis of Energy Intensity Developments in Manufacturing Sectors in Industrialized and Developing Countries', *Energy Policy*, **29**, 769–75.

Milgrom, P., Y. Qian and J. Roberts (1991), 'Complementarities, Momentum, and the Evolution of Modern Manufacturing', *American Economic Review*, **81**, 84–8.

Miller, S.M. and M.P. Upadhyay (2002), 'Total Factor Productivity and the Convergence Hypothesis', *Journal of Macroeconomics*, **24**, 267–86.

Mokyr, J. (1990), *The Lever of Riches: Technological Creativity and Economic Progress*, Oxford: Oxford University Press.

Mokyr, J. (1992), 'Technological Inertia in Economic History', *Journal of Economic History*, **52**, 325–38.

Moretto, M. (2000), 'Irreversible Investment with Uncertainty and Strategic Behavior', *Economic Modelling*, **17**, 589–617.

Morovic, T., F.J. Gründing, F. Jäger, E. Jochem, W. Mannsbart, H. Poppke, M. Schön and I. Tötsch (1987), *Energy Conservation Indicators*, Berlin: Springer Verlag.

Morovic, T., G. Gerritse, G. Jaeckel, E. Jochem, W. Mannsbart, H. Poppke and B. Witt (1989), *Energy Conservation Indicators II*, Berlin: Springer Verlag.

Mulder, P. (2003), *Description Sectoral Database OECD with Energy*, CPB Memorandum, The Hague: CPB Netherlands Bureau for Economic Policy Research.

Mulder, P. and J.C.J.M. van den Bergh (2001), 'Evolutionary Economic Theories of Sustainable Development', *Growth and Change*, **32**, 110–34.

Mulder, P., H.L.F. de Groot and M.W. Hofkes (2001), 'Economic Growth and Technological Change. A Comparison of Insights from a Neo-Classical and Evolutionary Perspective', *Technological Forecasting and Social Change*, **68**, 110–34.

Mulder, P., H.L.F. de Groot and M.W. Hofkes (2003), 'Explaining Slow Diffusion of Energy-Saving Technologies: A Vintage Model with Returns to Diversity and Learning-by-Using', *Resource and Energy Economics*, **25**, 105–26.

Nelson, R.R. (1994), 'What Has Been the Matter with Neoclassical Growth Theory?', in G. Silverberg and L. Soete (eds), *The Economics of Growth and Technical Change*, Aldershot: Edward Elgar, pp. 290–324.

Nelson, R.R. (1995), 'Recent Evolutionary Theorizing About Economic Change', *Journal of Economic Literature*, **33**, 48–90.

Nelson, R.R. and S.G. Winter (1974), 'Neo-classical vs. Evolutionary Theories of Economic Growth', *Economic Journal*, **84**, 886–905.

Nelson, R.R. and S.G. Winter (1977), 'In Search of Useful Theory of Innovation', *Research Policy*, **6**, 36–7.

Nelson, R.R. and S.G. Winter (1982), *An Evolutionary Theory of Economic Change*, Cambridge, MA: Harvard University Press.

Nelson, R.R. and S.G. Winter (2002), 'Evolutionary Theorizing in Economics, *Journal of Economic Perspectives*, **16**, 23–46.

Newell, R.G., A.B. Jaffe and R.N. Stavins (1999), 'The Induced Innovation Hypothesis and Energy-Saving Technological Change', *Quarterly Journal of Economics*, **114**, 941–75.

Nijkamp, P. and J. Poot (1998), 'Spatial Perspectives on New Theories of Economic Growth', *Annals of Regional Science*, **32**, 7–37.

Nijkamp, P. and A. Reggiani (1998), *The Economics of Complex Spatial Systems*, Amsterdam: Elsevier.

Nordhaus, W.D. (1967), 'The Optimal Rate and Direction of Technical Change', in K. Shell (ed.), *Essays on the Theory of Optimal Economic Growth*. Cambridge, MA: MIT Press, pp. 53–66.

Nordhaus, W.D. (2002), 'Modeling Induced Innovation in Climate Change Policy', in A. Grübler, N. Naki•enovi• and W.D. Nordhaus (eds), *Modelling Induced Innovation in Climate Change Policy*, Washington, DC: Resources for the Future, pp. 259–90.

OECD (1998), *ISDB: International Sectoral Database Description*, Paris: OECD.

OECD (1999), *Taking Action Against Climate Change: The Kyoto Proposal*, Paris: OECD.

OECD/IEA (2000), *Experience Curves for Technology Policy*, Paris: OECD/IEA.

Olson, M. (1982), *The Rise and Decline of Nations*, New Haven: Yale University Press.

Parente, S.L. (1994), 'Technology Adoption, Learning-by-Doing, and Economic Growth', *Journal of Economic Theory*, **63**, 246–369.

Park, S.H., B. Dissmann and K.Y. Nam (1993), 'A Cross-Country Decomposition Analysis of Manufacturing Energy Consumption', *Energy*, **18**, 843–58.

Payson, S. (1995), 'Product Evolution: What It Is and How It Can be Measured', *Eastern Economic Journal*, **21**, 247–62.

Peretto, P.F. (1996), 'Sunk Costs, Market Structure and Growth', *International Economic Review*, **37**, 895–923.

Phelps, E.S. (1966), 'Models of Technical Progress and the Golden Rule', *Review of Economic Studies*, **33**, 133–45.

Phylipsen, G.J.M., K. Blok and E. Worrell (1997), 'International Comparisons of Energy Efficiency–Methodologies for the Manufacturing Industry', *Energy Policy*, **25**, 715–25.

Phylipsen, G.J.M., K. Blok and E. Worrell (1998), *Handbook on International Comparisons of Energy Efficiency in the Manufacturing Industry*, Utrecht: Dept. of Science, Technology and Society at Utrecht University.

Pilat, D. (1996), *Labour Productivity Levels in OECD Countries: Estimates for Manufacturing and Selected Service Sectors*, OECD Economics Department Working Paper no 169, Paris: OECD.

Pindyck, R.S. (1991), 'Irreversibility, Uncertainty and Investment', *Journal of Economic Literature*, **29**, 1110–52.

Pindyck, R.S. (2000), 'Irreversibilities and the Timing of Environmental Policy', *Resource and Energy Economics*, **22**, 233–59.

Pindyck, R.S. (2002), 'Optimal Timing Problems in Environmental Economics', *Journal of Economic Dynamics and Control*, **26**, 1677–97.

Prywes, M. (1986), 'A Nested CES Approach to Capital–Energy Substitution', *Energy Economics*, **8**, 22–8.

Purvis, A., W.G. Boggess, C.B. Moss and J. Holt (1995), 'Technology Adoption Decisions Under Irreversibility and Uncertainty: An Ex Ante Approach', *American Journal of Agricultural Economics*, **77**, 541–51.

Quah, D. (1993), 'Galton's Fallacy and Test of the Convergence Hypothesis', *Scandinavian Journal of Economics*, **95**, 427–43.

Quah, D. (1997), 'Empirics for Growth and Distribution: Polarization, Stratification, and Convergence Clubs', *Journal of Economic Growth*, **2**, 27–59.

Ramirez, A., M. Patel and K. Blok (2002), *An Analysis of the Dutch Service Sector: Energy Consumption and Trends*, Utrecht: Copernicus Institute, Department of Science, Technology and Society, Utrecht University.

Rebelo, S. (1991), 'Long-Run Policy Analysis and Long-Run Growth', *Journal of Political Economy*, **99**, 500–521.

Reinganum, J.F. (1981), 'On the Diffusion of New Technology: A Game Theoretic Approach', *Review of Economic Studies*, **48**, 395–405.

Resource and Energy Economics (2003), *Endogenous Technical Change in Environmental Macroeconomics, Resource and Energy Economics* Special Issue, **25**, 1–126.

Romer, P.M. (1986), 'Increasing Returns and Long-Run Growth', *Journal of Political Economy*, **94**, 1002–37.

Romer, P.M. (1987), 'Growth Based on Increasing Returns due to Specialization', *American Economic Review*, **77**, 56–62.

Romer, P.M. (1990), 'Endogenous Technological Change', *Journal of Political Economy*, **98**, S71–S102.

Rosenberg, N. (1976), *Perspectives on Technology*, Cambridge: Cambridge University Press.

Rosenberg, N. (1982), *Inside the Black Box: Technology and Economics*, Cambridge: Cambridge University Press.

Rosenberg, N. (1994), 'Energy-Efficient Technologies: Past and Future Perspectives', in: N. Rosenberg (ed.), *Exploring the Black Box: Technology, Economics and History*, Cambridge: Cambridge University Press, pp. 161–89.

Ruttan, V.W. (2001), *Technology, Growth, and Development. An Induced Innovation Perspective*, Oxford: Oxford University Press.

Sahal, D. (1981), 'Alternative Conceptions of Technology', *Research Policy*, **10**, 2–24.

Salter, W.E.G. (1960), *Productivity and Technical Change*, Cambridge: Cambridge University Press.

Schipper, L. and S. Meyers (1992), *Energy Efficiency and Human Activity: Past Trends and Future Prospects*, Cambridge: Cambridge University Press.

Schipper, L., S. Murtishaw and F. Unander (2001), 'International Comparisons of Sectoral Carbon Dioxide Emissions Using a Cross-Country Decomposition Technique', *Energy Journal*, **22**, 769–78.

Schumpeter, J.A. (1934), *The Theory of Economic Development*, Cambridge, MA: Harvard University Press.

Shell, K. (1966), 'Toward a Theory of Inventive Activity and Capital Accumulation', *American Economic Review*, **56**, 62–8.

Shell, K. (1967), 'A Model of Inventive Activity and Capital Accumulation', in K. Shell (ed.), *Essays on the Theory of Optimal Growth*, Cambridge MA: MIT Press, pp. 67–85.

Silverberg, G. (1997), *Evolutionary Modeling in Economics: Recent History and Immediate Prospects*, MERIT Research Memorandum 97–013, Maastricht: MERIT.

Silverberg, G. and D. Lehnert (1993), 'Long Waves and "Evolutionary Chaos" in a Simple Schumpeterian Model of Embodied Technical Change', *Structural Change and Economic Dynamics*, **4**, 9–37.

Silverberg, G. and D. Lehnert (1994), 'Growth Fluctuations in an Evolutionary Model of Creative Destruction', in G. Silverberg and L. Soete (eds), *The Economics of Growth and Technical Change*, Aldershot, UK and Brookfield, US: Edward Elgar, pp. 74–109.

Silverberg, G. and B. Verspagen (1994a), 'Learning, Innovation and Economic Growth: A Long-Run Model of Industrial Dynamics', *Industrial and Corporate Change*, **3**, 199–223.

Silverberg, G. and B. Verspagen (1994b), 'Collective Learning, Innovation and Growth in a Boundedly Rational, Evolutionary World', *Journal of Evolutionary Economics*, **4**, 207–26.

Silverberg, G. and B. Verspagen (1998), 'Economic Growth and Economic Evolution: A Modeling Perspective', in F. Schweitzer and G. Silverberg (eds), *Evolution and Self-Organization in Economics*, Berlin: Duncker & Humblot, 239–64.

Silverberg, G., G. Dosi and L. Orsenigo (1988), 'Innovation, Diversity and Diffusion: A Self-Organisation Model', *Economic Journal*, **98**, 1032–54.

Simon, H.A. (1955), 'A Behavioral Model of Rational Choice', *Quarterly Journal of Economics*, **69**, 99–118.

Slaughter, S. (1993), 'Innovation and Learning During Implementation: A Comparison of User and Manufacturing Innovations', *Research Policy*, **22**, 81–95.

Smulders, S. and Th. v.d. Klundert (1995), 'Imperfect Competition, Concentration and Growth with Firm-Specific R&D', *European Economic Review*, **39**, 139–60.

Smulders, S. and M. de Nooij (2003), 'The Impact of Energy Conservation on Technology and Economic Growth', *Resource and Energy Economics*, **25**, 59–79.

Soete, L. and R. Turner (1984), 'Technology Diffusion and the Rate of Technical Change', *Economic Journal*, **94**, 612–23.

Solow, J.L. (1987), 'The Capital Energy Complementarity Debate Revisited', *American Economic Review*, **77**, 605–14.

Solow, R.M. (1956), 'A Contribution to the Theory of Economic Growth', *Quarterly Journal of Economics*, **70**, 65–94.

Solow, R.M. (1957), 'Technical Change and the Aggregate Production Function', *Review of Economics and Statistics*, **39**, 312–20.

Solow, R.M. (1994), 'Perspectives on Growth Theory', *Journal of Economic Perspectives*, **8**, 45–54.

Sørensen, A. (2001), 'Comparing Apples to Oranges: Productivity Convergence and Measurement Across Industries and Countries: Comment' *American Economic Review*, **91**, 1160–67.

Stern, N. (1991), 'The Determinants of Economic Growth', *Economic Journal*, **101**, 122–33.

Stokey, N.L. (1988), 'Learning by Doing and the Introduction of New Goods', *Journal of Political Economy*, **96**, 701–17.

Stoneman, P. (1983), *The Economic Analysis of Technological Change*, Oxford: Oxford University Press.

Stoneman, P. (2001), *The Economics of Technological Diffusion*, Oxford: Blackwell Publishers.

Stoneman, P.K. and P.A. David (1984), 'Adoption Subsidies vs. Information Provision as Instruments of Technology Policy', *Economic Journal*, **96**, 142–50.

Swan, T.W. (1956), 'Economic Growth and Capital Accumulation', *Economic Record*, **32**, 334–61.

Syrquin, M. (1984), 'Resource Allocation and Productivity Growth', in M. Syrquin, L. Taylor and L.E. Westphal (eds), *Economic Structure Performance – Essays in Honor of Hollis B. Chenery*, Orlando, FL: Academic Press, pp. 75–101.

Taheri, A.A. and R. Stevenson (2002), 'Energy Price, Environmental Policy, and Technological Bias', *Energy Journal*, **32**, 85–107.

Tondl, G. (2001), *Convergence after Divergence? Regional Growth in Europe*, Vienna and New York: Springer.

Torvanger, A. (1991), 'Manufacturing Sector Carbon Dioxide Emissions in Nine OECD Countries: 1973–87', *Energy Economics*, **13**, 168–86.

Unander, F., S. Karbuz, L. Schipper, M. Khrushch and M. Ting (1999), 'Manufacturing Energy Use in OECD Countries: Decomposition of Long Term Trends', *Energy Policy*, **27**, 769–78.

United Nations (2001), *United Nations Statistical Yearbook*, New York: United Nations.

Usher, A.P. (1954), *History of Mechanical Inventions* (1st edn 1929), Cambridge, MA: Harvard University Press.

Veblen, T. (1898), 'Why is Economics not an Evolutionary Science?', *Quarterly Journal of Economics*, **12**, 373–97.

Velthuijsen, J.W. (1995), *Determinants of Investment in Energy Conservation*, PhD Thesis, Amsterdam: SEO, University of Amsterdam.

Verspagen, B. (1992), 'Endogenous Innovation in Neo-classical Growth-Models: A Survey', *Journal of Macroeconomics*, **14**, 631–662.

Verspagen, B. (1993), *Uneven Growth between Interdependent Economies. The Evolutionary Dynamics of Growth and Technology*, Aldershot: Avebury.

Von der Fehr, N-H.M. and L. Sandsbråten (1997), 'Water on Fire: Gains from Electricity Trade', *Scandinavian Journal of Economics*, **99**, 281–97.

Vries, F.P. de (2003), *Environmental Policy and Technology Diffusion under Imperfect Competition. An Evolutionary Game Theoretical Approach*, PhD Thesis, Groningen: Groningen University.

Vuuren, D. van (1996), *Karakterisering ICARUS–3 Maatregelen ten behoeve van het CPB Energiemodel*, Report 96007, Utrecht: Department of Science, Technology and Society, University of Utrecht.

Wagner, K. and B. van Ark (1996), *International Productivity Differences: Measurement and Explanation*, Contributions to Economic Analysis (Vol. 233), Amsterdam: Elsevier.

Weyant, J. and T. Olavson (1999), 'Issues in Modeling Induced Technological Change in Energy, Environment, and Climate Policy', *Environmental Modeling and Assessment*, **5**, 67–85.

Worrell, E., L. Price, N. Martin, J. Farla and R. Schaeffer (1997), 'Energy Intensity in the Iron and Steel Industry: A Comparison of Physical and Economic Indices', *Energy Policy*, **25**, 727–44.

Young, A. (1993a), 'Invention and Bounded Learning by Doing', *Journal of Political Economy*, **101**, 443–72.

Young, A. (1993b), 'Substitution and Complementarity in Endogenous Innovation', *Quarterly Journal of Economics*, **108**, 775–807.

Young, A. (1994), 'Lessons from the East-Asian NICs: A Contrarian View', *European Economic Review*, **38**, 964–73.

Zettelmeyer, F. and P. Stoneman (1993), 'Testing Alternative Models of New Product Diffusion', *Economics of Innovation and New Technology*, **2**, 283–308.

Zwaan, B.C.C. van der, R. Gerlagh, G. Klaassen and L. Schrattenholzer (2002), 'Endogenous Technological Change in Climate Change Modeling', *Energy Economics*, **24**, 1–19.

Index